W9-CGY-190

REASONING WITH STATISTICS

REASONING WITH STATISTICS

FREDERICK WILLIAMS

Annenberg School of Communications
University of Southern California

SECOND EDITION

New York   Chicago   San Francisco   Atlanta   Dallas   Montreal   Toronto   London   Sydney   **HOLT, RINEHART AND WINSTON**

Library of Congress Cataloging in Publication Data

Williams, Frederick
  Reasoning with statistics.

  Includes bibliographies and index.
  1. Statistics.   2. Social sciences — Statistical methods.   I.   Title.
HA29. W525   1979   519.5   78–15744

ISBN 0–03–019536–5

Copyright © 1979, 1968 by Holt, Rinehart and Winston
All rights reserved
Printed in the United States of America

9 0 1 2   090   9 8 7 6 5 4 3 2 1

PREFACE

The second edition of *Reasoning with Statistics,* like the first edition, is intended for people who need to be knowledgeable readers of quantitative research literature in their fields but who lack the necessary statistical background. The main goal is to aid readers in understanding studies involving statistical methods rather than to prepare them to use such methods. Because the first edition, which contained mainly examples of communication research, also found significant use by persons in education, the second edition has a number of revised and new research examples which should be of interest to persons in both fields.

Most of the materials in this book evolved originally from class "handouts" or lecture materials that proved understandable to students who had no background whatsoever in statistics, including those students who expressed a particular lack of confidence in their mathematical backgrounds. Now, after a decade of use, we know that these materials are understandable to a wide variety of students in research seminars, courses that introduce research methods, and can be useful to supplement a main text in statistics classes.

We have retained the handbook features that proved so popular in the first edition, including lists of statistical symbols and key terms inside the front and back covers, setting key definitions apart from context, and including more than the usual detail in the index. All of this, we know, has made *Reasoning with Statistics* a useful resource book for people unaccustomed to reading reports of quantitative research.

Additionally, the chapters have been arranged and interrelated in order to provide continuity ranging from basic concepts (Part One), to statistical methods used for difference analysis (Part Two), methods used for relationships analysis (Part Three), and a new section on multivariate analysis (Part Four). The introduction of multivariate statistics into this second edition reflects the increase in their use in communication and education research in the last decade plus the anticipation that this use will continue to increase.

In reducing statistical concepts and applications to an understandable and useful level, the hazards of oversimplification have been weighed against the objective of aiding students to become more intelligent readers, rather than mainly users, of statistical methods. On the other hand, if you wish to *use* statistical methods, then you will need to read more than *Reasoning with Statistics*. Numerous suggestions for such readings are presented throughout the book.

The goal of *Reasoning with Statistics* will be close to fulfilled if it whets the appetite of a reader to undertake formal studies of statistical methods or increases the fidelity of the dialogue between those who swear by and those who swear at statistical methods. But the goal will be totally fulfilled if readers who have not had the time to study statistics will find quantitative research articles understandable and useful to them. (And this will require some help from those who write such articles!)

Many of my memories of the first edition bring to mind the witty conversations of the late Caleb Smith, the H.R.W. field representative who encouraged me to write my first book. Now, nine books later, I realized even more the contribution of such professionals in their field. I am grateful to the Literary Executor of the late Sir Ronald A. Fisher, F.R.S., to Dr. Frank Yates, F.R.S. and to Longman Group Ltd., London, for permission to reprint columns from Table III from their book *Statistical Tables for Biological, Agricultural and Medical Research* (6th edition, 1974). For their comment, I wish to thank S. G. Mueller; Raymond G. Smith, Indiana University; Ron Short, Whitworth College; Ernest Weiler, University of Cincinnati; Glen M. Broom, University of Wisconsin; Ronald S. Geizer, University of Minnesota; Sjef van den Berg, University of Connecticut; Richard J. Smith, University of Northern Colorado; Linda J. Shipley, University of Missouri; Steven L. Epstein, Columbia University; Donald J. Cegala, Ohio State University; Peter R. Monge, Michigan State University; Kenneth D. Frandsen, Pennsylvania State University; and George Rodman, Brooklyn College.

I must newly acknowledge in this second edition all of the insightful comments, criticism or praise, received from so many colleagues in communication and education over the last decade. I hope the second edition serves us all well.

FREDERICK WILLIAMS

Los Angeles, California
January 1979

CONTENTS

REASONING WITH STATISTICS

BASIC CONCEPTS

PART 1

STATISTICS AND RESEARCH

1

This is not a how-to-do-it book on statistics. Rather, this is more of a how-they-do-it book, an introduction to the practical logic and application of the statistical methods used in communications and education research. If you want to do statistics, this book will be useful only as a general orientation. There are many references in the following chapters for books on how to employ and to calculate statistical methods.

The reason for this book is a purely practical one. Students in communication and education often have backgrounds that do not include an introduction to statistical methods. Increasingly, however, they are expected to be intelligent readers of quantitative research reported in the textbooks and scholarly journals of their various areas. The aim of *Reasoning with Statistics* is to serve this type of reader. This viewpoint should answer the first question typically raised by a person faced with an introduction to statistics: "Do I need some special mathematical background or talent?" The answer is *no*—at least not for the present approach.

THE ROLES OF STATISTICS

Why do we reason with statistics? One response to this inquiry begins by turning to a brief introduction to the so-called scientific method used in studying human behavior.

We can do this best by viewing the scientific method in terms of its most evident characteristics. For one thing, this approach places stress on gaining knowledge through the process of observation. Whatever is said about behavior is reasoned from systematic observation and is tested and retested by observation. In other words, the so-called scientific approach attempts to anchor knowledge in terms of the physical reality it purports to explain.

Another characteristic of the scientific approach is that it uses rigorous procedures for gaining knowledge. We can see the reflection of these procedures in the typical outline of the report of scientific research.

1. *Problem:* a precise statement of what knowledge was sought and why it was sought.

2. *Method:* the plan of the research, that is, how the knowledge was gained.

3. *Results:* a precise statement of the knowledge that was gained.

If these steps seem formal, they are meant to be, for they stress the systematization of the scientific approach.

How, then, does statistics enter this picture? Perhaps we can best begin with the following definition.

> 1.1 *Statistics* (singular): a branch of applied mathematics that specializes in procedures for describing and reasoning from observations.

Practically speaking, we can think of statistics as providing tools used in carrying out the kind of observational research that we have described. The nature of these tools is mathematical—it draws its substance from the theory of numbers, their operations, inter-relations, and generalizations.

Observation and Statistics

One way of seeing how statistics enters the research picture is to consider the relationship between what it is that the researcher observes and how statistics serves as a means for reporting these observations. Consider first what is meant by a *phenomenon*.

> 1.2 *Phenomenon:* any object or event, the characteristics of which are susceptible to observation.

In the broadest sense, we use the word phenomenon to designate whatever segment of reality is under observation. For example, we might be concerned with the effects of a certain type of newspaper article on a certain type of reader. We could designate this overall situation as a phenomenon itself, or we might choose to divide it into particular phenomena, for example, characteristics of the article on the one hand, and characteristics of the readers' reactions on the other.

Given the designation of what is under study, the concern is usually with how characteristics of phenomena vary and how these phenomena affect other phenomena. When we talk about the varying characteristics of phenomena, we use the term *variable*.

> 1.3 *Variable:* an observable characteristic of an object or event that can be described according to some well-defined classification or measurement scheme.

In terms of the newspaper example, we might define articles as a variable according to the topic of the news items, for instance, local, regional, national, and international. Readers' reactions, as a variable, might be defined in terms of ratings of "interestingness." Presumably, then, a study could be concerned with the relationship between the variable of news topic and the variable of interestingness.

Whereas a phenomenon or variable is what is being observed, the researcher's reports of observations are considered as *data*.

1.4 *Data* (plural): the reports of observations of variables.

A characteristic feature of a study involving statistics is that the data are always in the form of some type of *measurement*.

1.5 *Measurement:* a scheme for the assignment of numbers or symbols to specify differing characteristics of a variable.

Suppose that interestingness of the newspaper articles were rated on a three-point scale, where *1* indicated high interest, *2* indicated average interest, and *3* indicated low interest. Such a scale would constitute the type of measurement employed in the study. Reports of observations, the data, would be in terms of these numbers.

Thus far we have not considered who would be providing the ratings of interestingness. That is, who or what would be one focus for measurement? Usually we are concerned with what is called a *population*.

1.6 *Population:* any class of phenomena arbitrarily defined on the basis of its unique and observable characteristics.

Notice that the above definition refers to phenomena, rather than strictly to people. In statistics, the word *population* is not limited to people; it refers to the total class of whatever is observed as a part of a study. If, for example, we were concerned with the manufacture of light bulbs, we could refer to a day's production as a population, if we so desired. In terms of the newspaper study, however, our population providing ratings is people. But this population includes only some well-defined group of people. The population, for example, might be the subscribers to the particular newspaper being studied. Depending on the researcher's interests, the population might be only female readers. As noted in the foregoing definition, whatever is defined as the research population is purely arbitrary; it depends on the limits of the research problem.

In many cases it is not feasible to measure an entire population; hence, the researcher settles for some portion of the population, called a *sample*.

1.7 *Sample:* a collection of phenomena so selected as to represent some well-defined population.

In the case of the light bulbs, a sample of the population might be every tenth bulb to come off the production line. For the newspaper study, the sample might be the names of 100 subscribers selected randomly from a list of the 1000 subscribers to the newspaper.

Once observation and measurement have been undertaken, what we have is a body of measurement data. Each datum is in numeric form and represents one observation about one item or member of a population. The numeric report itself tells us what particular characteristic we have observed in terms of our measurement scheme. The total body of data represents measures that we have taken upon every member of a sample of, in some cases, every member of a population.

Reasoning and Statistics

Thus far we have seen statistics mainly with reference to data, that is, particular numbers that we have assigned to observations. The main role of statistics comes in reasoning from these measurement data to the overall statements that we wish to make about the data. In this sense, statistics provides us with mathematical models for reasoning.

Suppose that the newspaper study required describing the average rating of interestingness of each of the four different types of newspaper articles. In this case we would probably use a familiar statistical procedure called the *arithmetic mean*. We can state this procedure in terms of its formula.

$$\text{Mean} = \frac{\Sigma X}{N}$$

The X in this formula is a symbol for a numeric value of a variable; it can be *any* numeric value. In the present case we would take it to represent the various ratings of interestingness that have been reported. The N is a symbol for another numeric value—the number of observations that we shall consider in determining the mean. In short, we identify X and N in the formula with variables that represent our defnitions of variables in the real world. Assuming that these identifications are adequate, we then follow the operations expressed in the formula—summation (symbolized by Σ) of the values of X, and division of this summed value by N to calculate the value of the arithmetic mean. The kinds of operations that we have followed are deductive in nature; they reflect mathematical rules.

The use of statistics in reasoning from data is in all cases somewhat the same as we have described for the arithmetic mean. Statistics provides us with a great variety of procedures, each with a pattern of mathematical deductions that provides some type of statistical conclusion. As was illustrated for the arithmetic mean, we identify variables that we have observed in the real world with variables that are symbolized in the formulas. Given a set of adequate identifications, we then reason according to the deductive operations of the formula until we reach some desired statistical result. Obtaining this result, we then reverse the process of identification and go from what we have calculated in the formula to some statement about the real world. For example, if the arithmetic mean of the interestingness of local news items is 2.5, we would then use this value to describe the average interestingness of this type of item to the population that was studied.

Even as simple a statistical procedure as the arithmetic mean can serve as a reminder of what is provided when reasoning with statistics, that is, reasoning mathematically from data in order to make some statement about the data. Sometimes we apply the word *statistic* to refer to the kind of "result" that a statistical formula provides. Thus, for example, statistically derived values that reflect an average, values that represent dispersion of measurements, or values that represent other such characteristics are called *descriptive statistics*.

> 1.8 *Descriptive statistics:* calculated values that represent certain overall characteristics of a body of data.

If the body of data comprises an entire population, such descriptive statistics, of course, describe that population. However, if the data comprise only a sample, then further statistics are necessary in order to talk about a population. These statistics are called *sampling statistics*.

> 1.9 *Sampling statistics:* calculated values that represent the probable deviations of sample characteristics from population characteristics.

Suppose again that the mean of the ratings of interestingness were 2.5. If we had measured the entire population, this would be the mean of that population. But if this were only the mean of a sample, we would have to employ a sampling statistic as an estimate of the likely position of the population mean. Such a statistic, when calculated, might tell us that if we considered the population mean as falling between 2.3 and 2.7, we would be risking only about 5 percent chance of error.

In sum, when we consider statistics in terms of reasoning from observations, there are two roles. First, there are statistics that provide descriptive statements about bodies of data. Second, there are statistics that provide bases for estimating population characteristics, based on our knowledge of sample characteristics. In many kinds of research, these two types of statistics go hand in hand.

THE RESEARCH PLAN

Another way of viewing the practical roles of statistics is to view them within the context of the research plan itself. Earlier we noted that research procedures were typically reported in terms of the *problem, method,* and *results.* Let us look at each of these aspects of the research plan in more detail and see how statistics may relate to each.

The Problem

Research typically begins with some need for information. Generally speaking, this need is expressed as a rationale leading up to the particular problem, that is, why a particular problem is worthy of study. The problem itself is then defined in a statement of the problem. It might be expressed as either a *purpose* or as a *question;* for example:

> The purpose of this study is to determine the average amount of television viewing of all freshman students at . . . university,
>
> or
>
> What is the average amount of television viewing of all freshman students at . . . university?

Another type of problem statement is a *hypothesis,* a statement susceptible to testing by reasoning from observations. For example:

> Freshman students engage in more television viewing than do sophomore students at . . . university.

A problem is stated in hypothesis form only when there are sufficient reasons to make a prediction, for instance, a prediction based on theory, a prediction based on prior observations, or both.

The statement of a given problem always has an effect on the type of statistics to be used in carrying out the study. Consider, for example, the preceding problem statements. In all of them, some particular type of measurement is implied; that is, viewing time will be defined in terms of some classification scheme. It might be in terms of hours per day or programs per day. Also, explicit in the purpose and question statements is the concept of *average.* This means that some statistically derived group value will necessarily be computed. This need for such a group value is implied in the hypothesis statement; it might be either an average or a total.

Notice also that the problem statement also provides a preliminary definition of the population to be studied. This too will have consequences upon the statistics to be used. Perhaps the researcher will measure entire populations—and will need only descriptive statistics—or the researcher may sample from the population and thus need both descriptive and sampling statistics.

The Method

Usually it is convenient to consider the method in terms of (1) the particular type of research method; (2) the subjects (persons being studied), materials, and procedures; and (3) the statistical analyses.

1. Method Types. Solving the research problem first involves an overall plan for gathering data. From a broad viewpoint, such plans fall into two main categories. The distinguishing feature is whether observations are conducted without attempting to manipulate the variables under study, or whether the researcher imposes particular manipulations upon the variables being studied, then observes the consequences. These two approaches are defined as follows.

> 1.10 *Descriptive method:* a research plan undertaken to define the characteristics or relationships, or both, among variables based on systematic observation of these variables.
>
> 1.11 *Experimental method:* a research plan undertaken to test relationships among variables based on systematic observation of variables that are manipulated by the researcher.

The descriptive method, or research involving observation but no direct control of variables, is also classified under a variety of other labels such as, for example, *empirical, survey, normative, analytic,* or even *clinical.* Some typical types of descriptive research in the area of communications include broadcast ratings, readership surveys, studies of message content, and public opinion polling. Descriptive research in education could involve assessment of curriculum needs, the use of new course materials, teachers' attitudes toward integration, and the like. The key feature in all such research is that some existing situation is being studied.

The experimental method, by contrast, is involved only when the researcher is actually manipulating the variable or variables under study. The usual intent is to test a hypothesis of cause and effect. That is, do manipulations on one variable lead to consequences on another variable? Customarily, the two variables are distinguished as the *independent* and *dependent* variables, respectively.

> 1.12 *Independent variable:* a phenomenon that is manipulated by the researcher and that is predicted to have an effect on another phenomenon.
>
> 1.13 *Dependent variable:* a phenomenon that is affected by the researcher's manipulation of another phenomenon.

Suppose, for example, that we were to test the hypothesis:

> Students taking English I via instructional television make better grades in the course than students taking a nontelevised version of the course.

If we were to employ the experimental method to test this hypothesis, part of the research strategy would entail setting up the two versions of the English I course as an independent variable. The effect to be studied, or the dependent variable, would constitute the final grades in the courses—presumably measured the same way for both sections. In other words, the study would center around the presumed causal relationship between the way a course is taught (independent variable) and its consequences on the final grades in the course (dependent variable).

Whether a study is descriptive or experimental in design has a number of consequences on the statistical models to be employed. In a descriptive study we would, of course, expect to find descriptive statistics being employed, and, if sampling is involved, these would be augmented by the use of sampling statistics. The experimental method, on the other hand, involves a slightly different consideration in terms of how the two types of statistics are employed.

In carrying out the experiment described above, we would select students for the televised course and the nontelevised course from *the same population.* That is, prior to the administration of the courses as an independent variable, we would assume that the students in both classes had the same potential for achieving a particular distribution of grades in the course. This is a crucial assumption, since if the students in the two courses differ after taking their respective classes, we would want to reason that it was due to the way the courses were taught, rather than the fact that the two groups differed in the first place.

After the administration of the courses, the dependent variable is assessed and the question is asked: Do these two groups of students now represent the same population in terms of grades in English I? Descriptive statistics will provide bases for characterizing the two groups in terms of average grade and dispersion of grades. But in this case we shall have a special use for sampling statistics. They will provide a basis for deducing whether it is likely that the two groups, as measured after having taken the course, still represent the same population in terms of English I grades. If it is found likely that the two groups do not now represent the same population, we have a basis for concluding that the courses affected them differently. Much more will be said on this particular use of statistics in a subsequent chapter on hypothesis testing (Chapter 5).

2. Subjects, Materials, and Procedures. We now are concerned with who is measured, what tools are used for measurement and for conducting the study, and precisely how the tools are applied to the population being studied.

In most types of communication studies, measurement centers on each individual as a member of a population being studied. Typically, we call this person a *subject,* usually abbreviated as S. From a statistical standpoint, we are interested in Ss because, as a minimal unit of observation, the number of such Ss defines at least the minimal number of observations made in a particular study. We shall need to know this number in order to calculate most descriptive statistics, and it further enters either directly or indirectly into our calculation of the various sampling statistics. We identify the number of observations included in a study as the variable N which is included in statistical formulas.

The *materials* of a study include whatever tools the researcher has employed in order to carry out the investigation. In an experiment, materials include whatever was used to manipulate the independent variable, for instance, the materials used in the televised and non-televised versions of English I. Also, we often think of materials in terms of the particular measurement instrument used for the study. For example, in studying television viewing time, the materials might include some type of questionnaire. Or, in the experiment on English I, the materials would include the final examination, which provides the measure of the dependent variable. From a statistical standpoint, the most usual concern with the materials of a study is the type of measurement that has been involved. Since this is the topic of Chapter 2, suffice it to say that materials may offer a variety of different types of measurement, and that the choice of particular statistics depends in part on the type of measurement used.

The *procedures* of a study refer to the precise manner in which the materials have been applied to the Ss and how this had led to the data of the study. Particular procedures carry implications concerning the use of statistics in a study. One example would be a case when repeated measurements are made on the same Ss; that is, each S serves as a basis for more than one observation in a study. In such a case, it is necessary to incorporate this information into selection and use of some of the different types of statistical procedures that might be brought to bear.

3. Statistical Analyses. Part of the task of designing a study involves deciding beforehand just what statistical procedures will be used and what criteria will be involved in reasoning from the statistical results back to the population under study. In a more conceptual sense, these decisions reflect the kinds of identification to be made in moving from variables being observed in the real world to variables symbolized in the statistical models and how, given the statistical results, identification will be made from the model itself back to the variables in the population. The selection of statistical models depends, of course, on what kinds of mathematical deductions the research plan calls for. These involve consideration, among other things, of what population characteristics are of interest (for example, averages, dispersions, and so on), what population comparisons might be made, and what type of measurement scale is involved.

Although they will be discussed in greater detail in Chapters 4 and

5, there are often criteria that must be established in deciding how to interpret the statistical results of a study. Recall, for example, what was said earlier in terms of risking only 5 percent chance of error when estimating the location of a population mean. Ideally, the tolerance for such error is a decision that should be made as a part of the original design of a study. In other words, in setting up the study, the researcher would plan to estimate the population mean within limits associated with only 5 percent chance of error—or the researcher might not want to risk as much error and thus set some other level, depending on the nature of the study. This type of decision also plays a key role in interpreting the results of an experiment. Recall the question posed in interpreting the results of the experiment that was described: Do the two groups, after undertaking different versions of the English I course, represent the same population in terms of grades in the course? The statistical results in this case will have an associated level of error. Part of the researcher's task in designing the experiment is to state how the likelihood of error accompanying statistical results will be interpreted in terms of the conclusions about the experiment. We shall see in Chapter 5 how this probability of error is incorporated into the logic of hypothesis testing.

A final consideration of the statistical analyses are the calculations themselves. In a practical sense, these refer simply to undertaking whatever calculations are defined by the statistical formula. In a more conceptual sense, these refer to how one follows the deductive mathematical rules and operations of the statistical model.

The Results

Generally speaking, the results of a study begin with whatever has been deduced by use of the statistical procedures. The task is then to interpret these statistical results in terms of what we identify them with in the population under investigation.

In terms of the role of statistics, the consequences of two types of logical patterns are reflected in the results of a study. On the one hand, the statistical results are themselves a product of a mathematical deduction. But when these results are identified with the real world, what we have in terms of the overall research strategy is a result that has been obtained inductively. That is, we have reached some generalization by means of observing a variety of particular instances of the variable under study—an inductive pattern. But in accomplishing this overall inductive pattern we have employed mathematical deductions to aid us in the process.

This foregoing distinction in patterns of reasoning is important when evaluating the results of a given piece of research. As will be seen in subsequent chapters, statistics provides rigorous basis for reasoning from data to what we have called *statistical results*. If phenomena in the real world have been adequately identified within the statistical model, and if the model has been used properly, we can have confidence in statistical conclusions. However, the researcher still has the burden of reasoning from the statistical results back to the real world. The criteria for evaluating this type of reasoning refers to the overall plan for research—its inductive adequacy—rather than to particular statistical models. Put into more practical terms, there are different criteria for evaluating the statistical aspect of a study, as compared with the overall plan for research. A knowledgeable reader of this type of research will understand just how reasoning with statistics is only one part—but a vital part—of an overall research plan.

SUMMARY

We can think of *statistics,* a branch of applied mathematics, as providing tools that aid in conducting scientific research. One aspect of statistics concerns how observations are reported in terms of *measurement.* The major aspect of statistics, however, incorporates procedures for reasoning mathematically from numerical data to whatever conclusions comprise the goal of a study. Such procedures are often classified into categories of *descriptive* or *sampling* statistics. Statistics can also be viewed within the context of an overall research plan, that is, in terms of the *problem,* the *method,* and the *results,* of which they form a part.

SUPPLEMENTARY READINGS

Campbell, Donald T., and Julian C. Stanley, *Experimental and Quasi-experimental Designs for Research.* Chicago: Rand McNally, 1963. This brief paperback originally a chapter, is a classic in the field; excellent for considerations of reliability and validity.

Isaac, Stephen, and William B. Michael, *Handbook in Research and Evaluation.* San Diego, Calif.: Robert R. Knapp, Pub., 1971. This paperback is excellent for a wide range of reference purposes.

Kerlinger, Fred N., *Foundations of Behavioral Research,* 2nd ed. New York: Holt, Rinehart and Winston, 1973. Part 1, Chapters 1–3, treats the "scientific" approach. Very readable.

Selltiz, Claire, Lawrence S. Wrightsman, and Stuart W. Cook, *Research Methods in Social Relations,* 3rd ed. New York: Holt, Rinehart and Winston, 1976. Good for the sociological or social psychological view.

MEASUREMENTS

2

We have already discussed in Chapter 1 how measurement serves the researcher as a numerical report of observations. In this respect, measurement bridges the gap between what a researcher reports as an observation of a variable in the real world and what has been defined as a variable in a statistical model. Reconsider the definition of measurement given in the last chapter—a scheme for the assignment of numbers to specify different characteristics of a variable. From a more detailed standpoint, it is necessary to understand the different types of particular measurement schemes, or what we call scales.

> 2.1 *Scale:* a specific scheme for assigning numbers or symbols to designate characteristics of a variable.

MEASUREMENT SCALES

Different measurement scales offer varying degrees of exactness in describing given characteristics. The system of symbols in measurement will usually comprise numerals, but what these numerals signify may range from the simple identification of categories for classification of characteristics to numbers representing "true values" of observed characteristics. Because scales range from low to high in the amount of information signified, it is common to speak of such scales in terms of levels of measurement. Understanding differences in these levels is important for two main reasons: (1) Selection of measurement scales depends on the type of information the researcher desires to record. (2) The scale that one selects may place restrictions on what models for statistical reasoning may be applied. The four types of scales available to the researcher are usually called the *nominal, ordinal, interval,* and *ratio.*

Nominal Scale

> 2.2 *Nominal scale:* the assignment of numbers or symbols for the purpose of designating subclasses that represent unique characteristics.

Sometimes called the *classificatory* scale, the nominal scale is the weakest level of measurement. Of the four types of scales, it signifies the least information about observations. When researchers classify observations into mutually exclusive categories, as in dividing color of eyes (a class) into such subclasses as blue, brown, and green, they are using a nominal scale. There is no intention to signify any order among the categories—as, for example, that blue eyes are of greater importance than brown eyes along some specified dimension. Different classes of observations might be identified in terms of arbitrarily assigned numbers, as in blue = 1, brown = 2, and green = 3 (or blue = 3, brown = 1, green = 2, and so on). The prime characteristic of such categories is that all observations assigned to a given category are equivalent in terms of some characteristic, and that they differ from phenomena in other designated categories in terms of this characteristic.

The foregoing phenomena, when measured in terms of a nominal scale, are often called *categorial* variables.

Numbers also come into our use of nominal scales when we count the frequency of observations assigned to the subclasses. This use of numbers, however, reflects the operation of enumeration, not the operation of nominal scaling. The operation of nominal scaling is simply the dividing of a given class of characteristics (for instance, eye color) into mutually exclusive subclasses (such as blue, brown, and so on).

Studies employing "content analysis" in communications or education research often employ nominal scales. For example, there have been a number of studies of the frequency with which different ethnic groups are depicted in children's textbooks. Nominal classification (or scaling) has been used to designate such ethnic categories as White, Black, Hispanic, Asian, and Native American. Then enumeration is employed in counting the frequency of which the characters in textbooks are classified into the ethnic categories. Eventually the relative frequency of characters in ethnic categories in the textbook are compared with the relative percentages of ethnic groups represented in the student population.

Ordinal Scale

2.3 *Ordinal scale:* the assignment of numbers or symbols for the purpose of identifying ordered relations of some characteristic, the order having unspecified intervals.

A first point in describing the qualities of an ordinal scale is to note that it incorporates the classificatory quality of the weaker nominal scale. The important point in ordinal scaling, however, is that among the subclasses there is an interrelationship of rank ordering. That is, every subclass can be compared with every other subclass in terms of some "greater than" (or "less than") relationship. The nature of the greater than depends, of course, on the characteristic under study. It may indicate such relations as more preferred, more emotional, more ethical, and so forth. What ordinal scaling does not represent is the magnitude of difference between ordered categories.

Questionnaires used in survey research often involve ordinal scaling. In a variety of studies you will find such questions as:

My preschool child watches television:
    (1) very much.
    (2) a little.
    (3) not very much.
    (4) not at all.

Please rank in order below the importance to you of the following media for keeping up with local news (radio, newspapers, television, magazines):
    1. _____
    2. _____
    3. _____
    4. _____

In both of the above cases numbers are assigned to indicate the relative order of something but there is no assumption that the difference between, say, 1.0 and 2.0 is equal to the difference between 3.0 and 4.0. Nor is there an assumption that the difference between 2.0 and 4.0 is twice the magnitude of the difference between 1.0 and 3.0. In short, nothing is said or assumed about the magnitude of intervals between numbers other than that they are unspecified. But it *is* assumed that "2" is less than "3", or "6" is greater than "5", and so on. The numbers are labels of order, nothing more.

Interval and Ratio Scales

2.4 *Interval scale:* the assignment of numbers for the purpose of identifying ordered relations of some characteristic, the order having arbitrarily assigned and equal intervals but an arbitrary zero point.

2.5 *Ratio scale:* the assignment of numbers for the purpose of identifying ordered relations of some characteristic, the order having arbitrarily assigned and equal intervals, but an absolute zero point.

The first characteristic to note in the definitions of interval and ratio scales is that magnitude is treated in terms of known and equal intervals. A second characteristic of these scales is that they are thoroughly quantitative; they are always expressed in numbers, or something representing numbers. It is also useful to point out that interval scales meet most, and ratio scales meet all, of the mathematical assumptions necessary to perform arithmetic operations. By contrast, there are substantial restrictions on which arithmetic operations can be applied when using nominal or ordinal scales.

There are, however, some important differences between interval and ratio scales. These differences stem from the consequences of having an arbitrary zero point (interval scale) as compared with a true zero point (ratio scale). When considering an interval scale, the focus is on the *difference* between values on the scale. Because the numbers are based on an arbitrary zero, their values, as such, may not be adequately reflective of the real world, nor are they isomorphic to the structure of arithmetic. However, these numbers do adequately define intervals on the interval scale; therefore, the magnitudes of difference along an interval scale provide useful designations of characteristics of the real world. Further, these differences are compatible with the rules of arithmetic. Consider, for example, the two interval scales illustrated in Figure 2.1. These are the Fahrenheit and the centigrade scales for measuring temperature. Both are interval scales; zero is arbitrarily defined. The true zero would necessarily represent that point at which the characteristic being measured vanishes, that is, the absolute absence of heat. The locations of arbitrary zero of the Fahrenheit and centigrade scales were set out for convenience and, further, because the nature of a true zero in temperature was unknown at the time.[1]

To illustrate how we can deal with differences, but not intrinsic values, in using the interval scale to measure phenomena, consider the following: The ratio of two differences on the Fahrenheit scale should equal the ratio of equivalent differences on the centigrade scale.

Suppose that we set this ratio as the differences between two-thirds and one-third of the range in temperature between freezing

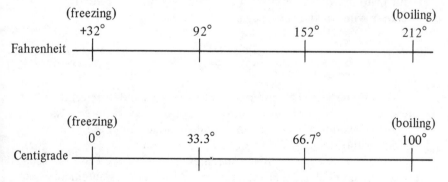

FIGURE 2.1   Fahrenheit and centigrade temperature scales

[1] Zero on the Fahrenheit scale approximates the temperature produced by mixing equal quantities (by weight) of snow and common salt; on the centigrade scale, zero is the freezing point of water.

and the boiling point of water. If the earlier statement is true, then the following should be true:

$$\frac{212°F - 92°F}{212°F - 152°F} = \frac{100°C - 33.3°C}{100°C - 66.7°C}$$

Which, indeed, it is (2 = 2). What, then, about ratios of *values?* Would ratios of values representing the two-thirds and one-thirds scale points between freezing and boiling be equal across the two interval scales? The answer is no — a consequence of arbitrary zero points.[2]

$$\frac{152°F}{92°F} \neq \frac{66.7°C}{33.3°C}$$

Consider next the two ratio scales, inches and centimeters. Suppose that the line in Figure 2.2 is exactly one yard (36″ or 91.44 cm) in length. With a ratio scale, ratios of scale values as well as differences reflect the real world. Take the example of ratios of two-thirds to one-third differences; here we shall consider a yard as the overall range, similar to the range in temperature between freezing and boiling. What was true for differences on the interval scale is true for the ratio scale; for example:

$$\frac{36″ - 12″}{36″ - 24″} = \frac{91.44 \text{ cm} - 30.48 \text{ cm}}{91.44 \text{ cm} - 60.96 \text{ cm}}$$

But in terms of ratios of values, what was not true for the interval scale is true for the ratio scale:

$$\frac{24″}{12″} = \frac{60.96 \text{ cm}}{30.48 \text{ cm}}$$

To summarize: The interval scale specifies the magnitude of differences among ordered points on a scale. Such intervals, rather than the absolute values of the numbers that define them, are applicable to the rules of arithmetic and to statements about phenomena. The ratio scale, by contrast, specifies values as well as differences that are applicable to arithmetic operations and to statements about phenomena.

In cases where the research problem calls for more than classification or ranking, it would, of course, be desirable to work with ratio scales. In enumeration (counting) operations, we do usually work

---

[2] The symbol $\neq$ is read as "not equal to."

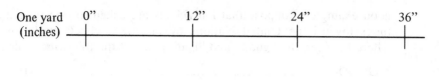

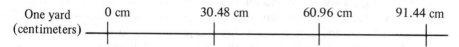

FIGURE 2.2   Inch and centimeter scales

with a ratio scale. Zero is usually absolute, and numeric values are usually isomorphic with the rules of arithmetic. We say usually because sometimes the enumeration operation is applied to psychological tests where, for example, we are counting the number right. Solely from the standpoint of enumeration, zero right on such a test is an absolute value. However, from a psychological standpoint, zero right may not characterize zero ability. In fact, the notion of zero ability may never lend itself to feasible or even to useful definition. In such cases, then, we are considering scaling rather than strict enumeration, and we are back with the properties of an interval or an ordinal scale.

In communications research, ratio scales often come into use in the measurement of physical phenomena—characteristics of objects, events, or biophysical characteristics of the human organism. A biophysical example is represented in a respirometer, which provides measures of air moved (cubic centimeters) during any specified time period. Such a measure has an absolute zero (that is, no air moved) and is comparable with alternative scales also based on this zero point (for example, cubic inches).

Unfortunately, psychological aspects of behavior do not usually lend themselves to ratio scaling. For one thing, a true zero is often difficult to conceptualize, and is perhaps not even useful in characterizing psychological characteristics. For example: What does zero intelligence mean? At what point does a state of encoder anxiety truly vanish? There do, however, exist psychological attributes, such as in studies of psychophysical scaling (for instance, sound perception), where properties of a ratio scale are achieved. But when measuring most psychological aspects of behavior, the communications or education researcher is typically faced with use of an interval scale and, more often than we would like, an ordinal scale.

The fine line between ordinal and interval scaling is a recurring problem in measurements such as those used to describe psychological judgments. The frequently employed "semantic differential" is a good example.[3] Suppose that individuals are asked to rate their attitudes toward the United Nations by placing a check mark somewhere between the "good" and "bad" poles of the following scale:

GOOD _____ : _____ : _____ : _____ : _____ : _____ : _____ BAD

A conservative assumption is that the scale has the properties of ordinality. That is, for example, a check mark two cells to the left of the center would indicate a greater association between good and United Nations than a check mark one cell to the left of the center. The assumption of only ordinality, however, will quite drastically restrict the researcher in terms of statistical procedures applicable to the data. A more liberal assumption is that the scale cells have equal intervals, thus achieving the properties of an interval scale. What this involves is assuming that the distances between the cells are equal and that such equality is psychologically meaningful to the user. If this assumption can be made, then a far greater variety of statistical models are available for data analysis and interpretation. No doubt, this raises a dilemma for researchers. Should they assume only the power of an ordinal measurement and thus restrict the scope of the analyses, or should they assume interval measurement and thus gain the use of more powerful statistical procedures for reasoning from the data? The grounds for answering this question are not simple. Indeed, the answer usually depends on a compromise between what researchers want to achieve in terms of having confidence in the validity of results, and what they are willing to risk as error.

Scales of the ordinal, interval, and ratio types are associated with *continuous* variables.

MEASUREMENT ADEQUACY

Procedures for statistical reasoning are themselves without safeguards for warning the researcher when a measurement scale leads to spurious conclusions. Nothing in the procedures will prevent the researcher from using numbers from the ordinal scale in a statistic requiring an interval scale; the problem comes in reasoning back to something true about the real world. Accuracy in measurement is like accuracy in language — it depends on the user. Two facets of the measurement problem concern the qualities of *validity* and *reliability*.

Validity

2.6 *Validity:* the degree to which researchers measure what they claim to measure.

[3] Charles E. Osgood, George J. Suci, and Percy H. Tannenbaum, *The Measurement of Meaning*. Urbana: University of Illinois Press, 1957.

The question of validity is a question of "goodness of fit" between what the researcher has defined as a characteristic of a phenomenon and what he or she is reporting in the language of measurement. In more abstract terms, it is the question regarding the truth of the measurement. For example: To what degree do the scores on a listening comprehension test represent what we conceive to be a person's ability in listening? To what degree does the numerical coding of a check mark on a seven-cell scale represent a person's attitude? Questions such as these characterize the problem of validity of measures.

Reliability

2.7 *Reliability:* the external and internal consistency of measurement.

The problem of reliability raises a question somewhat different from validity. Rather than the truth of a measure, external reliability asks: If the measures were applied and reapplied under precise replication of conditions, would the same results be obtained? Closely related to this is the question raised by internal reliability: If the measure comprises subparts (as in a multiple-section test of some type), do these parts contribute to equivalent results? Reliability, then, poses an objective of consistency.

PROBLEMS OF VALIDITY AND RELIABILITY

The practical concern with validity and reliability can be illustrated in a study once conducted by the author. The research problem involved measuring the time taken by encoders who were responding to a signal to create a specified message. Both the signal used to start the encoding sequence and the signal indicating the encoder's completion of the message were recorded by pen markings on a paper tape calibrated in time intervals. Encoding time was defined as the duration between these start and completion signals.

What the author claimed to measure (validity) was the time taken to encode the message. From a technical standpoint, the assessment of validity required answers to such questions as: Were the start and completion signals accurately recorded as anticipated on the tape? Did the time calibration on the tape correspond to what is defined by known standards as seconds? From a more behavioral standpoint, however, were such questions as: Did the encoder recognize the start signal as anticipated? Was the time between the start and completion signals always indicative of encoding behavior (and not some other behavior, for instance, picking up a pencil that has dropped, and so on).

What the author hoped would be consistent (reliability) would be answered in response to such questions as: If the same encoding event could be precisely replicated (without practice effects), would the time measures be precisely the same?

The foregoing example also illustrates two further points—the general sources of criteria for evaluating validity and reliability, and the consequences that validity and reliability have on one another.

In order to assess validity of the time measures, it was necessary to use some type of outside standards, that is, the comparison of the measure with something other than itself. One part of this validity assessment depended on an observer verifying that the recorded signals for start and completion did, in fact, represent the events. Validation of the time measure required the use of some separate standards, in this case, a stop watch. The main test for reliability, however, was simply a comparison of the measure with itself. In this case, a series of constant start-to-completion durations was repeated a number of times, and the recordings compared. Although there are many and more complex approaches to assessing validity and reliability than this illustration provides, the ideas that evaluation of validity requires some type of outside standard, and that evaluation of reliability requires some way of comparing a measure with itself, remain basic considerations.

A final point concerning validity and reliability is the implication that they carry for one another. Does validity imply reliability? Does reliability imply validity? The answers are yes and no, respectively. If the signals of the encoding events were properly recorded on the tape and if the time durations corresponded to an outside criterion, such validity would imply (at least at the moment) that the measures are also reliable. A fallacious assumption is that reliability implies validity. The time calibration of the equipment may be perfectly consistent with itself, but this carries no implication whatsoever that it is measuring what has been defined as a *second*.

SUMMARY

Measurement involves the application of some numeric or symbolic scheme used to designate characteristics of a variable. As such, measurement bridges the gap between what we record as observations of a variable in the real world and what we may define as a variable in a statistical model. Different measurement schemes include: the *nominal*, a designation of subclasses of characteristics; the *ordinal*, subclassifications that are rank ordered; the *interval*, ranking with known intervals but with an arbitrary zero; and the *ratio*, ranking with known intervals but with a true zero. *Validity* of measurement is the extent to which the researcher measures what he or she claims to measure. *Reliability* refers to the internal and external consistency of measurement.

SUPPLEMENTARY READINGS

Kerlinger, Fred N., *Foundations of Behavioral Research,* 2nd ed. New York: Holt, Rinehart and Winston, 1973. Part 8 on "Measurement" is comprehensive and readable.
Siegel, Sidney, *Nonparametric Statistics for the Behavioral Sciences.* New York: McGraw-Hill, 1956. Old but not outdated; Chapter 3 presents a concise section on measurement "levels" and the implications for choice of statistical methods.

DISTRIBUTIONS

3

Once data have been recorded according to a measurement scheme (Chapter 2), the next step is to prepare some overall description of what has been observed. In any type of inductive research we are interested in making generalizations about what has been observed. Statistics provides us with ways by which to reason in a systematic and objective manner to various descriptions that represent generalizations about data. We then take these descriptions as a basis for saying something about the real world.

Descriptions of data may range from simply an arrangement of measurement values in some tabular or graphic form to the use of descriptive indexes based on certain statistical models. In one way or another, such descriptions or descriptive indexes are based on the *distribution* of measurements.

> 3.1 *Distribution:* a collection of measurements usually viewed in terms of the frequency with which observations are assigned to each category or point on a measurement scale.

TABULAR AND GRAPHIC DESCRIPTIONS

Nominal and Ordinal Measures

As indicated in Definition 3.1, the two main ingredients in a distribution are the measurement scale and the frequency of observations assigned to different categories or points on that scale. In some cases we are interested in distributions where the measurement scale is simply one of classification by category, that is, a nominal scale. Figure 3.1 shows a hypothetical example of a graphic summary of data where nominal scaling is involved. Here the categories represent messages on particular topics (for example, "cancer cures") and sources of these messages. The values listed under the heading N are tabular entries that indicate the number of people who were asked to rate the trustworthiness of the sources of the messages. The graphic portion (bar graph) indicates the percentage of these people (of N) who rated the message source as trustworthy. Notice that the measurement categories in Figure 3.1 could be arranged in any order. Also, each bar in the graph could be replaced with any type of drawing that could represent the percentages. We might even

Topic	Source	N	Percent rating source as trustworthy
Cancer cures	Medical Journal	200	95%
	News Tabloid	180	6%
Atomic power	TV Special	190	92%
	Politician	185	3%
Energy shortage	Business Magazine	200	80%
	Oil executive	205	17%
Future of Television	Textbook	200	90%
	Newspaper	205	20%

FIGURE 3.1   Ratings of credibility of message sources (simulated)

use quantities or portions of some type of cartoon figure to represent ratings of trustworthiness (that is, a pictograph).

Many times we are concerned with measures that represent a related series of categories along a scale. In contrast to the categories shown in Figure 3.1, it is now useful to depict the order that relates the measurement categories. Table 3.1 provides an example of a hypothetical distribution involving ordered categories. Here the measure is one of general attitude toward educational television (ETV); the scale ranges from a category labeled very negative to a category

Table 3.1   **Comparative Rating of General Attitude Toward ETV by Teachers With and Without ETV Available (Simulated)**

Rating	With ETV $N = 50$ (%)	Without ETV $N = 100$ (%)
Very positive	25	11
Moderately positive	23	35
Mildly positive	20	31
Neutral	18	10
Mildly negative	11	7
Moderately negative	3	5
Very negative	0	1

labeled very positive. Again the frequencies are reported in terms of percentages. This tabulation also shows how two distributions can be presented for comparative purposes; the distinction here is one of schools with or without ETV.

Interval and Ratio Measures

Whereas the foregoing distributions involved categories of observations, it is often the case that the distribution will involve a continuous scale of particular measurement values. In many types of psychological measures we refer to each particular measurement as a *score*. Rather than being interested in scores simply as categories or ordered categories, we are interested in them as points along a scale.

Suppose that we wanted to measure how much information was retained by persons who had been exposed to a particular message. For this purpose consider that we had constructed a 15-question test covering the message material. In particular terms, this measure can provide any of 16 scores ranging from 0 (all wrong answers) to 15 (all correct answers), and these scores could be depicted as a 16-point scale (counting 0 as a point). For example:

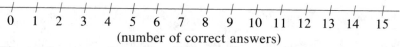

(number of correct answers)

The most commonly used symbol to indicate a single score or measurement value is X. What we would have, then, if the information-retention test were administered to 50 people, would be 50 identifications of X. Suppose that the following scores were obtained:

5	15	4	9	14	11	3	4	7	6
4	10	13	10	5	9	8	7	6	5
8	9	4	6	0	5	11	7	6	4
7	7	11	1	12	9	8	7	7	5
6	12	15	8	4	8	9	6	7	5

As presented in the above form, the scores comprise what is usually called ungrouped data. In this form they tell us little about the overall picture. We might find it convenient to group the data in tabular form; usually this is in terms of a frequency table, as in Table 3.2.

In graphic form, these same data might be presented as a *frequency polygon* or as a *histogram,* such as shown in Figure 3.2. Notice that the shape of the distributions approximates the distribution of tally marks in Table 3.2. What we have in either the frequency polygon or the histogram is a graphic presentation of the frequency that each score was observed. In the case of both presentations, we have plotted above each score or measurement interval a point that

Table 3.2   **Example of a Frequency Table**

Score	Tally	f
15	//	2
14	/	1
13	/	1
12	//	2
11	///	3
10	//	2
9	/////	5
8	/////	5
7	////////	8
6	//////	6
5	//////	6
4	//////	6
3	/	1
2		0
1	/	1
0	/	1

corresponds to the frequency with which the score was observed in the data. In the frequency polygon, we have simply connected the points, whereas in the histogram we have placed a horizontal line at each point to correspond to the interval taken by each score.

Shapes of Distributions

Simply by inspection, we can determine some of the characteristics of the overall collection of scores. For one thing, we can see (Fig. 3.2) that the most frequent scores are roughly scores from 4 to 9, and that relatively fewer persons had extremely high or low scores. Further, the gross shape of a frequency distribution is usually indicative of certain general characteristics. Consider, for example, distributions labeled A through E in Figure 3.3. Distribution A approximates the shape of the distribution we have been discussing. Compare this distribution with distribution B, which is more peaked than distribution A. This indicates that fewer individual differences were discriminated by the measuring instrument; that is, people were more alike in terms of the measurement. A distribution can also be flat, as in the case of distribution C. A flat distribution indicates more discrimination among individual differences in terms of the measures.

If a distribution has a peak that tends to be displaced at one or the other ends of the measurement scale and a tail that is strung out in the opposite direction, this pattern is called *skewedness*. It is described in terms of the direction of the stringing out of the tail of the curve. In distribution D, for example, the tail points toward the lower scores; it is called a negative skew. Distribution E shows just

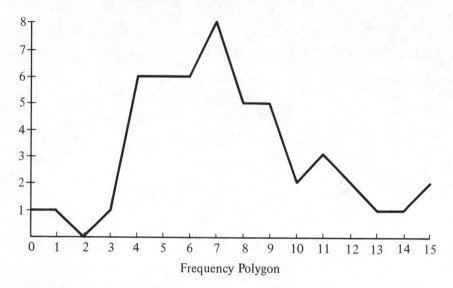

Frequency Polygon

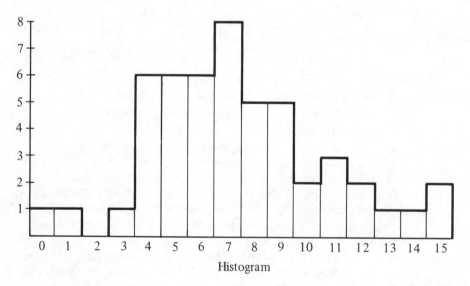

Histogram

FIGURE 3.2   Graphic displays

the opposite. Here the tail points toward the higher scores; it has a positive skew. These descriptions are sometimes confusing, since a negative skew means that the scores are clustered (the peak) in the positive direction; that is, people tend to correctly answer most of the questions on the test. A positive skew would indicate just the opposite – that people tended to do poorly on the test.

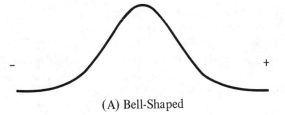

(A) Bell-Shaped

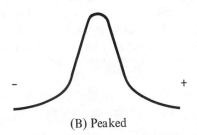

(B) Peaked

(C) Flat

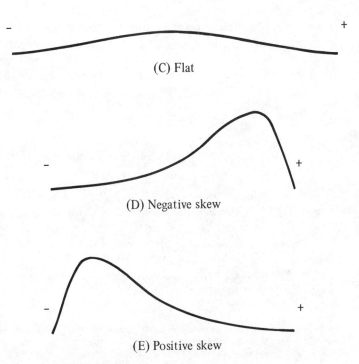

(D) Negative skew

(E) Positive skew

FIGURE 3.3   Shapes of distributions

No doubt, it would be cumbersome if it were always necessary to present tabular or graphic descriptions in order to report the characteristics of a collection of measures. Fortunately, we have more economical methods to report the nature of distributions. Such methods involve the use of statistical indexes that characterize particular aspects of measurement distributions.

INDEXES OF CENTRAL TENDENCY

One series of indexes tells us about the central tendency of the scores in a distribution. In simpler terms, these refer to how scores tend to cluster in a particular distribution. Central tendency is what we are refering to when talking about an average score. Three indexes are the *mode,* the *median,* and the *mean.*

The Mode

3.2 *Mode:* the most frequent score in a distribution.

Once the scores are grouped, the mode is determined by inspection. It is simply the score that has been tallied the most (Table 3.2) or the score under the peak of the distribution (Fig. 3.2). It is possible, however, to have a distribution in which every score occurs an equal number of times. In such a case there is no mode. Also, it is sometimes the case that a distribution has two or more scores that are of the highest yet equal frequencies. Such a distribution is considered as bimodal, trimodal, or multimodal, or as the case may be. All modes are then reported. For the data shown in Table 3.2 and Figure 3.2, the mode is 7. *Mo* is sometimes used as a symbol for the mode.

The Median

3.3 *Median:* the midpoint or midscore in a distribution.

Once scores are ordered by magnitude or are grouped, the median is defined as that point above which and below which one-half of the scores fall. For the data in Table 3.2 and Figure 3.2, the median is 7. Of the 50 scores, 21 are greater than 7, and 21 are less than 7. If we divided the eight scores that were of the value of 7, we could define 7.0 as the midpoint of the distribution. This would then give 25 above 7.0, and 25 below, or half above and half below.[1] A symbol sometimes used for the median is *Mdn.*

---

[1] If the scores within the median interval do not divide in equal groups, then the median point is determined by interpolation. Any elementary computational statistics book describes this procedure.

The Mean

3.4 *Mean:* the sum of the scores in a distribution divided by the number of scores.[2]

As an index of central tendency, the mean is by far the most important to us. It is the most sensitive index of central tendency of the types discussed, and it is a concept incorporated in most further statistical procedures to be discussed. Most of us have calculated this type of average score long before any encounter with statistics. In Table 3.2 and Figure 3.2, the sum of the scores is 369; this divided by 50, the number of scores, yields 7.38, which is the mean. A commonly used symbol to denote the mean is $\bar{X}$, where the bar over the X distinguishes the mean from a value indicating a particular individual's score. There are, however, additional symbols used to signify the mean. Usually the symbol M is used to refer to the mean of a sample, whereas the Greek lowercase letter *mu* (μ) is used to refer to the mean of a population.

It is useful for us to realize that the mean is sensitive to all of the scores in a distribution; we can think of it as a balance point. Consider for example, the following scores:

$$8 \quad 10 \quad 13 \quad 9 \quad 7 \quad 11 \quad 10 \quad 12 \quad 10 \quad 9 \quad 11$$

If we were to locate these scores each as a single weight along a bar divided according to the measurement scale, the balance point would be the mean as illustrated in Figure 3.4. Given the mean, we can view the scores in the distribution in terms of their deviations about the balance point. Here (Fig. 3.4) the sum of the negative deviations should equal the sum of the positive deviations. If we take their algebraic sign into account, the sum of the negative deviations should cancel out the sum of the positive deviations. The concept of deviations about the mean as a balance point is an important point in certain statistical procedures to be discussed subsequently.

Some Comparisons of the Mean, Mode, and Median

The mode and median are not necessarily balance points as is the mean. Take the scores in Figure 3.4, for example. To change the mode (10), we would have to reduce the number of scores in the category having the most frequent occurrence, or else add sufficient scores in another category to create a new mode. To change the median (10), we would have to add or shift scores so as to change the half-above, half-below divisions of the distribution. In both cases

---

[2] Strictly speaking, this is called the *arithmetic mean.*

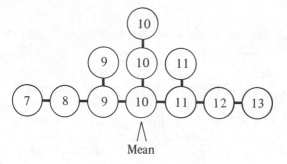

Scores about the mean

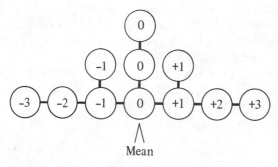

Score deviations about the mean

FIGURE 3.4   The mean as a balance point

there are many scores that could be changed without any effect on either the mode or the median.

If the mean is sensitive to all scores in a distribution, why consider using the mode or median as an index of central tendency? The most practical argument favoring the mode is that it is the easiest to obtain. Once the scores are grouped, if only by frequency of occurrence, the mode can be obtained simply by inspection. If the scores are ranked, the median is also easier to obtain than is the mean. There are, however, some further considerations when comparing the use of the median and the mean. Because of its characteristic as a balance point, the mean may sometimes be "pulled away" from what appears to be the point where the scores cluster in a distribution. This becomes the case when there are some extremely high or low scores in a distribution. Examine the distributions A, B, and C in Figure 3.5. Note that in distribution A, the mean is to the left of the median. This is because the few extremely low scores cause the balance point to shift to the left. Similarly, in distribution B, the mean is pulled toward the extremely high scores. The location

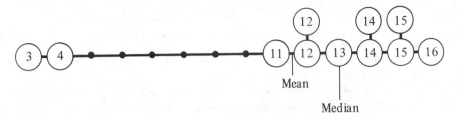

Mean

Median

(A) Mean biased toward negative skew

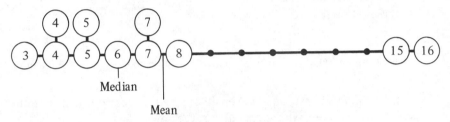

Median

Mean

(B) Mean biased toward positive skew

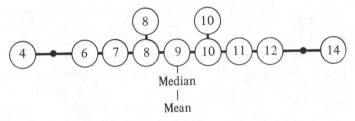

Median

Mean

(C) Mean and median are equal

FIGURE 3.5 Effects of skewedness upon relative locations of the mean and median

of the mean in both distributions A and B illustrates how extreme scores in a skewed distribution tend to bias the location of the mean. In such cases, the median is usually more representative of the central tendency of the distribution—or the average score, so to speak. When a distribution is symmetrical, as in distribution C, the mean and the median are the same.

INDEXES OF DISPERSION

Just as there are indexes that describe how scores tend to cluster about the center of a distribution, there are statistical indexes that describe the dispersion or scatter across the measurement scale. The most common of these indexes are the *range*, the *variance*, and the *standard deviation*.

The Range

3.5 *Range:* the highest score in a distribution minus the lowest score.[3]

The range is the crudest index of the dispersion of a distribution. Just as the mode and the median are insensitive to all but a few scores in a distribution, so is the range. In fact, it is based on only two scores, the highest and the lowest. For the distribution in Table 3.2 (or Fig. 3.2), the range is 15. The inclusion of any number of scores between the 0 and 15 points would have no effect on this index of dispersion. There is no particular symbol commonly used to denote the range.

The Variance

3.6 *Variance:* the mean of the squared deviation scores about the mean of a distribution.

Like the mean, variance is sensitive to all scores in a distribution. As defined above, variance reflects the deviations of scores about the mean of a distribution. These deviations are the same as we discussed earlier when considering the mean as a balance point among the scores in a distribution. Consider again the example of deviations (Fig. 3.4), but this time we shall square each deviation value, as in Figure 3.6. Notice first that all of the deviations in Figure 3.6 are positive; this is the result of squaring the original deviation values. We can talk about the general variability shown by these deviations simply by finding their sum (called *sum of squares;* in this case it is equal to 30). On the other hand, we want an index of dispersion that somehow indicates an average value of the squared deviations. We get this by simply calculating the arithmetic mean of the sum of squares; the result is *variance* (for example, 30 divided by 11, which equals 2.73 when rounded). A lowercase s with a squared sign, s^2, and the lowercase Greek sigma with a squared sign, σ^2, are the symbols usually used to denote variance. In a strict sense, s^2 refers to the variance of a sample, whereas σ^2 refers to variance of a population.

The Standard Deviation

3.7 *Standard deviation:* the square root of the mean of the squared deviation scores about the mean of a distribution; more simply, the square root of the variance.

Although Definition 3.7 may seem confusing at first, you will probably note that it is very similar to the definition of variance. In fact, it is much easier to define standard deviation as simply the

[3] Range is also sometimes defined as the *total range,* or the highest score minus the lowest score, plus 1.

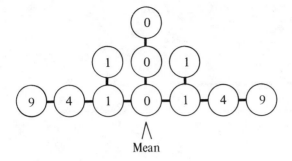

FIGURE 3.6   Squared deviations about the mean

square root of the variance. This relationship is consistent with the difference between symbols used to denote variance and symbols used to denote standard deviation. The symbol s (that is, the square root of s^2 refers to the standard deviation of a sample, and σ refers to the standard deviation of a population. In terms of the variance (2.73) that we were just discussing, the standard deviation would be equal to $\sqrt{2.73}$ or 1.652. For the data shown earlier in Table 3.2, the sum of squared deviations about the mean (7.38) is equal to 532 (rounded). This divided by the number of scores (50) in the distribution equals 10.64, or the variance. Finally, the square root of the variance, that is, the standard deviation, is equal to 3.26.

One practical advantage of using standard deviation rather than variance as an index of dispersion is that its values are smaller and easier to work with. Of more importance, however, is that standard deviation gives us a basis for estimating the probability of how frequently we can expect certain scores to occur upon the basis of sampling. We shall discuss this point in Chapter 4.

Interrelations Among Score Deviations, Variance, and Standard Deviation

Since deviations from the mean and the concepts of variance and standard deviation are crucial to the understanding of many statistical models, it is important to be able to conceptualize their interrelationship. This can be done geometrically. Considering the raw data shown as (A) in Figure 3.7, examine distribution (B) in this same figure. This distribution pictures deviations of these scores about the mean. Imagine these deviations in terms of linear distances (that is, lines 3 units long, 2 units long, and so on). Next consider what we have when the deviations are squared (C in Fig. 3.7). The total area of all of the squares is equal to what we have called the *sum of squares*. When we divide this total area by the number of scores in the distribution, we have the average square and its area is equal to the *variance* of the distribution. The area of the average square is shown as D in Figure 3.7. Finally, *standard deviation* can

(A) Raw scores: 7, 13, 8, 12, 10, 9, 9, 11, 10, 10, 11

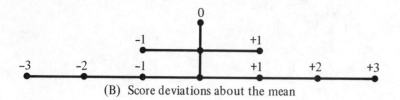

(B) Score deviations about the mean

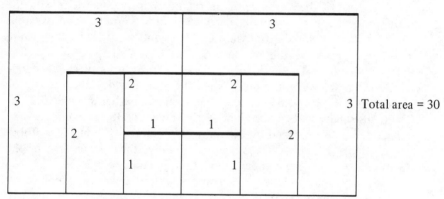

(C) Squared deviations about the mean

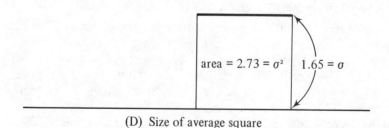

(D) Size of average square

FIGURE 3.7   Interrelations among deviations, variance, and standard deviation

be seen as the length of one side of the average square shown as D
in Figure 3.7. (Recall that the area of a square is equal to the length
of one of its sides squared.) This square provides a geometric con-
ception of the relationship between standard deviation and variance.

DISTRIBUTIONAL INDEXES AS DESCRIPTIVE STATISTICS

In Chapter 2 we saw how each observation could be reported in the language of measurement as data, whereas in the present chapter we saw how the collective characteristics of a given body of data could be reported in terms of selected descriptive indexes. These indexes are examples of *descriptive statistics*.

Such descriptive statistics, then, give us the capability of reasoning back to the real world in terms of what we have statistically reasoned from data. For example, we can consider statistical descriptions of a collection of data such as provided earlier in Table 3.2. If we wanted to describe the average level of information gain, we could turn to such descriptive indexes as the mean, the mode, or the median. Or if we were interested in how individuals differed in terms of this measurement, we could turn to the range, the variance, or standard deviation. All of these are examples of using descriptive statistics when reasoning with statistics.

SUMMARY

Distributions enable us to make collective reports about a body of measurements. Such reports might be presented in terms of graphic displays. Simply by inspection, graphic displays can be interpreted in terms of how the scores cluster about a particular point or how the scores tend to be dispersed along the measurement scale, or how they may be skewed. *Descriptive statistics* provide indexes for characterizing various aspects of distributions. Indexes of how scores tend to cluster, or central tendency, include the *mode, median,* and *mean.* Indexes of the scatter or dispersion of scores include the *range, variance,* and *standard deviation.*

SUPPLEMENTARY READINGS

Almost any basic statistics book will provide a guide for computation of the descriptive indexes discussed in the present chapter. However, the following provide various specialized treatments that may be of interest to you.

Kerlinger, Fred N., *Foundations of Behavioral Research,* 2nd ed. New York: Holt, Rinehart and Winston, 1973. Chapter 6 on "Variance" is an excellent introduction to a concept underlying many statistical methods.

Levin, Jack, *Elementary Statistics in Social Research,* 2nd ed. New York: Harper & Row, 1977. See Chapters 3, 4, 5, and 6; all quite comprehensive.

McCall, Robert B., *Fundamental Statistics for Psychology,* 2nd ed. New York: Harcourt Brace Jovanovich, 1975. Chapters 2 and 3 treat distributions. McCall also has a companion "Study Guide" and workbook, published separately.

Weinberg, George H., and John A. Schumaker, *Statistics: An Intuitive Approach,* 3rd ed. Monterey, Calif.: Brooks/Cole, 1974. Chapters 2, 3, and 8 are useful for materials on distributions. This book has an innovative approach to statistics, hence is a good "alternative" on most basic topics.

PARAMETERS

In the previous chapter we saw how distributions and descriptive indexes of distributions enabled us to report the collective characteristics of a given body of data. In all such cases, the descriptive statistics—for example, mean, variance, and so on—were applied only to the distribution at hand; there were no attempts to generalize beyond the data to some larger population. By contrast, in the present chapter we shall be concerned with the characteristics of populations.

INDEXES OF POPULATIONS

As discussed in Chapter 1, it is rarely feasible to observe and measure an entire population. What we do, instead, is to measure some portion (sample) that is taken as representative of a population. Then, in talking about the population, two types of statistical models are brought to bear. Descriptive statistics provide a basis for calculating such indexes as the mean, variance, and standard deviation of a sample. Given these indexes, we then apply sampling statistics to make estimates of how these indexes apply to a population. The essence of the problem, then, is to see how we can make statements about a population when we have observed only a sample of that population. This can be expressed more concisely in the terminology of statistics. Consider this special usage of the word *statistic:*

4.1 *Statistic* (singular): a characteristic of a sample.

In the above sense, a statistic would refer to a characteristic such as an arithmetic mean, the variance, or the standard deviation of a given sample. This usage distinguishes between what we call characteristics of samples (that is, statistics) and what we call *parameters*.

4.2 *Parameter:* a characteristic of a population.

Suppose, for example, that we calculated the average television-viewing time of a sample of freshman students at a given university; the purpose is to make an inference about the viewing time for all freshmen at that university. The arithmetic mean of the sample would be called a statistic; this statistic is then part of the basis for estimating a parameter, that is, the estimate of the mean viewing time for all the freshman students. The relationship between statistics and parameters underlies the following definition of *statistical inference:*

4.3 *Statistical inference:* the process of estimating parameters from statistics.

RANDOM SAMPLING

Thus far we have loosely considered a sample as a collection of observations that is somehow representative of a population. Let us now modify the definition of a sample given in Chapter 1.

4.4 *Random sample:* a collection of phenomena so selected that each phenomenon in the population had an equal chance of being selected.

Quite apart from any situation involving statistics, you have probably seen random sampling involved in many different types of situations. In almost any type of survey research, various strategies will be employed to try to ensure that every member in the defined population has an equal chance of being selected in the sample. Recall, too, that in the illustration of an experiment involving instructional television (Chapter 1), the two groups of subjects were drawn so that they initially would be from the same population, students in English I. This would entail random selection of the subjects from that population for the experiment, then further random assignment to one of the two groups.

In the simplest terms, we can say that if each phenomenon or unit in a population had an equal chance of inclusion in a sample, then that sample will have characteristics that can be used as a basis for estimating population characteristics. In slightly more specific terms, if random sampling is involved, this will enable us to draw upon a body of statistical theory that incorporates the mathematical relations between sample characteristics and population characteristics. Random sampling allows us to employ the logic of statistical inference, that is, estimating parameters from statistics.

In order to understand better how statistics relate to parameters, it will be helpful to see the relation conceptually. This entails understanding a number of different concepts pertaining to distributions.

THREE KINDS OF DISTRIBUTIONS

One way of viewing the logic of statistical inference is to see the relation among three types of distributions—the *sample distribution,* the *population distribution,* and what is called a *sampling distribution.*

Sample Distribution

> 4.5 *Sample distribution:* the frequency with which observations in a
> sample are assigned to each category or point on a measurement
> scale.

You should note that the above definition characterizes a particular
type of distribution within the more general definition of a distribution given earlier (Definition 3.1). Recall the array of scores presented in Table 3.2. Suppose that now we considered them as representing a random sample of a well-defined population. As a sample,
their sample distribution would be the same as the distribution presented in the foregoing chapter (Fig. 3.2), or we can view this same
distribution in Figure 4.1. Notice that in Figure 4.1 we can also
view certain characteristics of this sample distribution, that is, its
mean (*M*) and standard deviation (*s*). In Chapter 3 we saw how
these two statistics could be calculated. What we want to see next is
how these statistics relate to their respective parameters, that is, a
population mean and standard deviation.

Population Distribution

For purposes of illustration, suppose that we knew the precise characteristics of the population from which the foregoing sample was
drawn. We could plot these characteristics as a *population distribution,* such as shown in Figure 4.1.

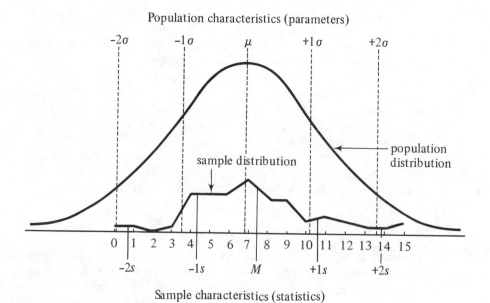

FIGURE 4.1   Comparison of sample and population distributions

4.6 *Population distribution:* the frequency with which all units or observations in a population would be assigned or expected in each category or point on a measurement scale.

Figure 4.1 illustrates a conceptual relation between sample distributions and populations distributions. What we usually observe is the sample distribution. This is obtained as a result of random sampling, taking of measurements, and calculating distribution characteristics. At the same time, we assume the existence of a population distribution having as yet unknown characteristics. Put into statistical terms: Given the knowledge of sample characteristics (statistics), our task is to estimate population characteristics (parameters). This is where the *sampling distribution* comes in.

Sampling Distribution

Thus far we have considered distributions involving the frequency of particular units or observations as they are assigned or expected at different points along a measurement scale. Suppose now that we considered the distribution of a statistic itself. To illustrate what we mean by this, examine Figure 4.2. Imagine that from a given population we have drawn a series of successive yet independent samples. That is, we have drawn a sample, noted its mean (M), then replaced it in the population, drawn another, noted its mean, replaced it, and so on. Our knowledge of the laws of chance (or the consequences of random sampling) tells us to expect some degree of variation among the means of samples drawn from the same population. Consider examples A, B, and C in Figure 4.2; these illustrate how sample means may vary among successive samples. For purposes of reference, we have imagined for the moment that we know the population distribution. What we have, then, in these three examples is an illustration of how sample distributions can vary within a given population distribution.

Now consider example D in Figure 4.2. Suppose that after drawing many samples, we plotted a distribution of the sample means. This is an example of a distribution of a statistic; it is also known as a *sampling distribution.*

4.7 *Sampling distribution:* the frequencies with which particular values of a statistic would be expected when sampling randomly from a given population.

In the present illustration we have what is known as the sampling distribution of a mean. Other statistics also have sampling distributions. We could, for example, consider the sampling distribution of s. Also, as will be seen in subsequent chapters, there are further statistics that characterize differences between or among samples, or relations among measures. We can conceive of each of these statistics as having a sampling distribution.

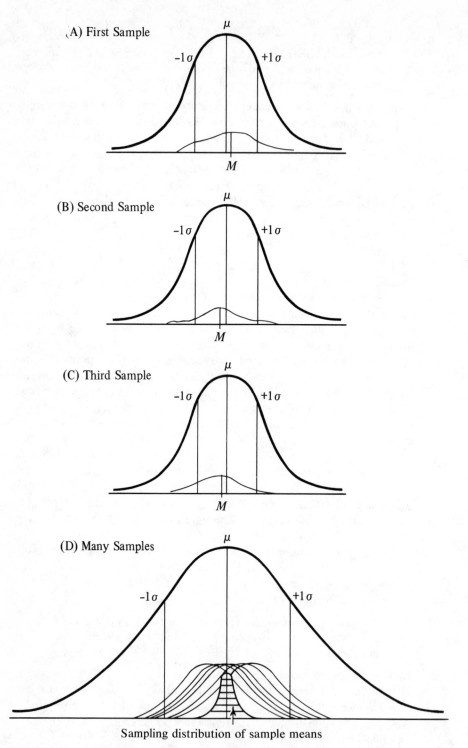

(A) First Sample

μ

-1σ $+1\sigma$

M

(B) Second Sample

μ

-1σ $+1\sigma$

M

(C) Third Sample

μ

-1σ $+1\sigma$

M

(D) Many Samples

μ

-1σ $+1\sigma$

Sampling distribution of sample means

FIGURE 4.2   Distribution of sample means as a statistic

DISTRIBUTIONS IN TERMS OF PROBABILITY

Another way of viewing the three kinds of distributions that we have discussed thus far is to see them in terms of probabilities rather than frequencies. The problem with a frequency distribution is that it is always tied to a particular number of observations and how this number is divided among the different categories or points along a scale. Suppose that instead of considering a distribution in terms of the frequencies of units at different categories or points, we considered each frequency in terms of what proportion of the total number of units it represented. Such proportions, then, could be generalized to cases where there are different total numbers of units involved. Moreover, to make the description even more useful, we consider these proportions as probabilities. This assumes that the total probability of units is considered to be equal to 1.0 in a distribution, with the distribution of these units expressed in terms of varying probabilities of occurrence at points or categories along a scale.

Comparison of a Frequency and a Probability Distribution

To use a simple example to illustrate the foregoing point, suppose that we were tossing six coins at a time and were interested in the number of heads that we would expect as the outcome of any given toss. First we will consider what frequencies would be involved. Using a mathematical basis[1] to determine what the laws of chance would predict if we tossed the six coins 64 times, we would have the distribution shown in Figure 4.3 – that is, no heads (or all tails) in 1 out of 64 tosses, or 1 head (5 tails) in 6 out of 64 tosses, or 2 heads (4 tails) in 15 out of 64 tosses, and so on.

As a general model of our expectations, this distribution of frequencies – that is, the particular numerical values – are based on a total of 64 tosses. To give the description of this distribution generality to cases involving other than a total of 64 tosses and to make this description more compatible to describing expectations, we can convert this description to probabilities. Thus, as also shown in Figure 4.3, we can say that the probability of obtaining an outcome of no heads is roughly .0156, one head is .0938, two heads is .2344, and so on. The sum of the probabilities of all categories of outcomes (analogous to categories on a scale) would be 1.0

The Normal Distribution Curve

4.8 *Normal distribution curve:* a definition of a particular functional relation between deviations about the mean of a distribution and the probability of these different deviations occurring.

[1] Called a *binomial expansion.*

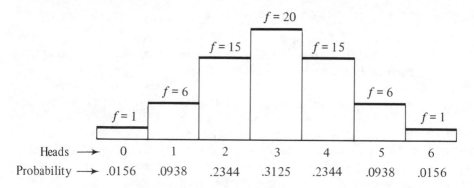

FIGURE 4.3   Distribution of the number of heads expected in tossing six coins 64 times: outcomes expressed in frequencies (*f*) and probabilities

The normal curve is a theoretical distribution only. We may never find sample distributions or population distributions that precisely fit the normal curve, but in many cases we do not go very far wrong if we assume that certain populations have this kind of distribution. Assuming that a population has this distribution then implies that a random sample of that population will also have a normal distribution. Additionally, the sampling distribution of certain statistics can also be assumed to have a normal distribution, for example, the sampling distribution of the mean in a normally distributed population.

A common illustration of the normal distribution curve is given in Figure 4.4. Notice its parameters—namely, a mean (μ) and a standard deviation (σ). In other words, the deviations in this distribution are expressed in σ units. Notice also how probability is incorporated into the area of this distribution. That is, the curve is divided into deviation segments, and the area above each represents some proportion of the area under the total curve. If we want to express these segments in terms of probability, we take the total area of the curve as equal to 1.0, then express each segment as either a proportion or a probability. For example, the probability of a deviation occurring between the mean (μ) and $+1\sigma$ is .3413, or between the mean -1σ is .3413, or between -1σ and $+1\sigma$ is .6826. It is the definition of how particular deviational segments (that is, 1σ, 2σ, 3σ, and so on) divide the area under this curve that expresses the "particular functional relation" mentioned in Definition 4.8. In other words, the particular division of the curve into these areas according to σ segments is the characteristic quality of the normal curve.

If you summed the area segments of the curve shown in Figure 4.4, you would note that they total only .9972, rather than 1.0. What this indicates is that there are further deviations from the mean

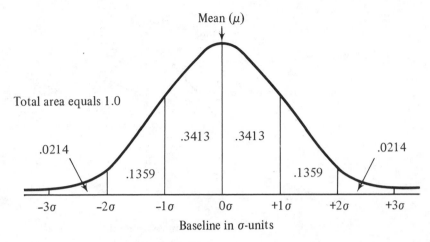

FIGURE 4.4   The normal distribution curve

(+4σ, −4σ, −5σ, and so on) that theoretically account for the re-
mainder of the area, or probabilities of occurrence. Being very extreme from the midpoint, these deviations have extremely low probabilities of occurrence. In fact, theoretically the tails of this curve
extend out to infinity.

   What we have, then, in the normal distribution curve is simply a
model of a distribution having certain relations between points along
the baseline (expressed in σ units) and probabilities of occurrence
(areas in segments of the curve).

The Normal Curve as a Population Distribution

   How do we relate the normal distribution model to actual measurement values? The tie-in comes when we identify the parameters of
the normal distribution with the parameters of some actual measurement distribution. You have probably noticed that the σ symbol is
the same as we used earlier for *standard deviation*. If we want to
talk about such deviations in terms of the theoretical curve, we define them as σ units. In other words, the standard deviation of the
theoretical curve is equal to 1.0. But if we want to talk about these
same deviations in terms of scores, we find out what σ is equal to in
terms of actual score units. Additionally, we assign a value of a
mean to the distribution. Suppose that we know the parameters of a
given population to be $\mu = 48$, $\sigma = 4.0$. Assuming that this population fits the normal curve model, then its parameters could be
shown as in Figure 4.5. In terms of actual score units, the mean (μ)
is now shown on the distribution as 48. That point at $+1\sigma$ above the
mean is now defined as 52 (given that $1\sigma = 4.0$) or -1σ is defined as
44, or -2σ is defined as 40, and so on. What we have done is to as-

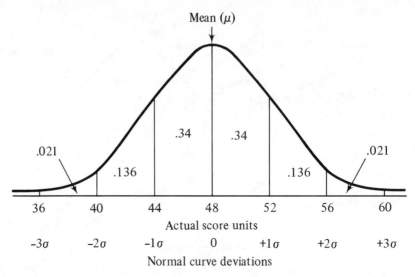

FIGURE 4.5   Actual scores in a distribution fitting the normal curve, where $\mu = 48$ and $\sigma = 4.0$

sign the parameters (μ and σ) of a known population to the parameters of the normal curve model. We can now interpret this curve (Fig. 4.5) in terms of the probabilities with which various scores would be expected. The following questions and answers illustrate the kinds of information that we can obtain from the curve.

1. What is the probability of obtaining a score between 48 and 52?
   Answer: Approximately .34. This is the same as defining the area between the mean and $+1\sigma$.
2. What is the probability of obtaining a score between 44 and 52?
   Answer: About .68, because this encompasses the area from -1σ to $+1\sigma$.
3. What is the probability of obtaining a score between 40 and 56?
   Answer: About .95, because this encompasses the area from -2σ to $+2\sigma$.

   Given the idea that we can assign different means and values of σ to the normal curve, it follows that this curve might not always be the same shape as shown in Figure 4.5. For example, as the value of each σ unit gets larger, we would expect the curve to spread out; or as each σ unit is smaller, the curve would tend to pull in and have more of a peak. This point is mentioned because it is sometimes erroneously assumed that a normal distribution is always bell shaped. The key feature of the normal distribution is the defined relation between each σ unit and respective areas under the curve.

The Normal Curve as a Sampling Distribution

Recall our statement earlier that we conceive of a sampling distribution as a distribution of a statistic, for example, sample means. In part D of Figure 4.2, we saw a distribution of sample means. We can take this distribution and view it in terms of the normal curve model. Examine Figure 4.6; this is a sampling distribution of the mean given in the previous example (Fig. 4.5). Notice that this distribution, like any other normal curve, has σ units. Only in this case each σ_M unit refers to the deviations of sample means about the population mean. We call the σ_M unit a *standard error of the mean;* it is a sampling statistic.

> 4.9 *Standard error of the mean:* the standard deviation of a distribution of sample means.

As we used the normal curve model as a basis for estimating probabilities of different scores occurring in a population, here we use the normal curve model as a basis for estimating the probabilities with which different sample means will occur. The following questions and answers illustrate the information that we can gain from this curve (Fig. 4.6).

1. What is the probability of a sample mean occurring between 48.0 and 48.4?
   Answer: Approximately .34. This is the area between the mean and $+1\sigma_M$.
2. What is the probability of obtaining a sample mean between 47.6 and 48.4?

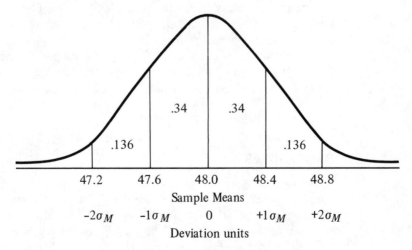

FIGURE 4.6   Sampling distribution of sample means where $M = 48$, and $\sigma_M = 0.4$

Answer: Approximately .68, because this encompasses the area from $-1\sigma_M$ to $+1\sigma_M$.

3. What is the probability of obtaining a sample mean between 47.2 and 48.8?

Answer: About .95, because this encompasses the area from $-2\sigma_M$ to $+2\sigma_M$.

Just as sample means have a sampling distribution, other statistics have their sampling distributions. We have centered upon the mean because it is the descriptive statistic most commonly considered in relation to a parameter.

SAMPLING ERROR

What we have discussed thus far has illustrated the relations among sample distributions, population distributions, and sampling distributions. For example, if we have sampled randomly from a normally distributed population, we can assume that certain relations hold between what we have found as characteristics of our sample and what the characteristics of the population are likely to be. One way of illustrating this relation is in terms of the sampling distribution of a statistic—for example, the mean. A more general way of thinking about what a sampling distribution of a statistic provides is in terms of *sampling error.*

4.10 *Sampling error:* an estimate of how statistics may be expected to deviate from parameters when sampling randomly from a given population.

In practical terms, whereas random sampling will yield sample characteristics that tend toward the population characteristics, we cannot expect sample characteristics to be precisely the same as population characteristics. The laws of chance allow sample characteristics to deviate from population characteristics, but our knowledge of these laws allows us to estimate what kinds of deviations to expect. This is the essence of statistical inference and the underlying logic of sampling statistics.

PRACTICAL PROCEDURES

In the case of actual research we begin only with the knowledge of sample characteristics. We can then incorporate the logic discussed in the preceding sections if we can make the assumption that we have sampled randomly from the population. Additionally, if we can meet the assumptions necessary for the normal curve model, then our procedures for estimating population characteristics will be based on the population and sampling distributions fitting that model.

Let us consider again the data presented in Figure 4.1 (Table 3.2), but this time we shall assume that we know nothing about the population, except that it is normally distributed. What kinds of statements might we make about population characteristics? Let us consider the population mean (μ) and the population standard deviation (σ).

Population Mean

We have already seen how the sampling statistic, *standard error of the mean* (σ_M) describes the likely deviations of sample means about the population mean. In making estimates of the population mean, we begin with the calculation of this statistic. One method for such calculation is shown in Table 4.1. Here the result is that $\sigma_M = .47$. Given this value, Figure 4.7 illustrates a sampling distribution of sample means estimated for this distribution, where $M = 7.38$. Thinking purely in terms of "best estimate," we could say that the mean of the sample is the best single estimate of the population mean. Usually, however, we take a more conservative approach that incorporates what we know about the sampling distribution of the sample means. Two such approaches, the first more conservative than the second, are as follows.

If we assume that the population mean lies between 6.17 and 8.59, there would be a probability of only .01 that we are wrong. Note in Figure 4.7 what this estimate is based on. The range be-

Table 4.1 **Examples of Calculations of Estimates of Parameters (Data from Table 3.2, Fig. 4.1)**

Sample Characteristics:

$$M = 7.38$$
$$N = 50$$
$$\Sigma x^2 = 532$$

$$s = \sqrt{\frac{\Sigma x^2}{N}} = \sqrt{\frac{532}{50}} = 3.26 \text{ (rounded)}$$

Population Characteristics:
Standard error of the mean (σ_M)

$$\sigma_M = \sqrt{\frac{\Sigma x^2}{N(N-1)}} = \sqrt{\frac{532}{50(50-1)}} = .47 \text{ (rounded)}$$

Standard deviation (σ)

$$\sigma = \sqrt{\frac{\Sigma x^2}{N-1}} = \sqrt{\frac{532}{50-1}} = 3.295 \text{ (rounded)}$$

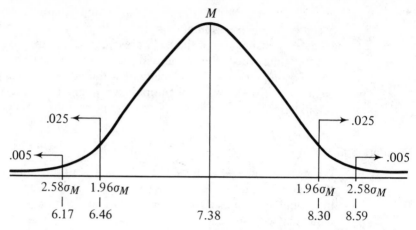

FIGURE 4.7   Sampling distribution where $M = 7.38$, $\sigma_M = .47$

tween 6.17 and 8.59 encompasses .99 of the area of the sampling distribution. This corresponds to the range between $-2.58\sigma_M$ to $+2.58\sigma_M$, which leaves an area of approximately .005 in each tail of the curve. In other words, we are incorporating estimates that are both above and below the mean of this distribution. If, by some unfortunate stroke of luck, we were indeed wrong, this would mean that the population mean is somewhere outside of the area between $-2.58\sigma_M$ to $+2.58\sigma_M$ on the curve in Figure 4.7. You can see, however, that as the population mean might be farther and farther away from the sample mean, its chances of being so are less and less probable.

If we were willing to assume a slightly greater chance of error in our estimation, we might assume that the population mean lies between 6.46 and 8.30 and say that there is a probability of .05 that we are wrong. This estimate is within the range of $-1.96\sigma_M$ to $+1.96\sigma_M$, which leaves approximately .025 area in each tail of the curve.

The foregoing estimates reflect the types of decisions we usually make when interpreting a sampling statistic. This actually entails an hypothesis-testing decision procedure, a topic treated in greater detail in the next chapter.

Population Standard Deviation

There are a number of alternative methods for estimating the standard deviation of a population, based on a knowledge of sample characteristics. One of these is presented in Table 4.1. Recall that in Chapter 3 we noted that the standard deviation of a distribution (without any implication of its being a sample) was calculated by the formula for *s* shown in Table 4.1. When a distribution is considered as a sample, its standard deviation is often calculated by use of the formula shown for σ in Table 4.1; as implied by the symbol σ, this is also taken as the best estimate of the population standard deviation. If, as with the population mean, we wanted to take a more conservative approach in making our estimate, we could plot a sampling distribution of sample standard deviations. This sampling distribution would have as its standard deviation a *standard error of the standard deviation* (σ_σ), a concept analogous to the standard error of the mean. We could then take this sampling distribution and make various statements about the location of the population standard deviation and accompany these statements with estimates of likely error.

SUMMARY

The essence of statistical inference is to reason from characteristics of samples (statistics) in order to estimate characteristics of populations (parameters). We are in a position to employ the logic of statistical inference when samples have been *randomly* drawn from populations. Sample characteristics are incorporated in a *sample distribution,* whereas a *population distribution* incorporates the population characteristics that are to be estimated. A *sampling distribution* characterizes the likely deviations of a given statistic about a parameter.

When distributions can be approximated to the *normal distribution curve,* we have at least one type of basis for making probability statements about parameters and likely sampling error of statistics. In a practical situation there are various methods for calculating estimates of population characteristics, given sample characteristics.

SUPPLEMENTARY READINGS

Kerlinger, Fred N., *Foundations of Behavioral Research,* 2nd ed. New York: Holt, Rinehart and Winston, 1973. Chapters 7 and 8 cover probability and sampling; Chapters 11 and 12 include materials on the relations between statistics and parameters.

Levin, Jack, *Elementary Statistics in Social Research,* 2nd ed. New York: Harper & Row, 1977. Chapter 6 covers the normal curve, Chapter 7 covers population characteristics.

Nie, Norman H., and others, *Statistical Package for the Social Sciences,* 2nd ed. New York: McGraw-Hill, 1975. Chapter 14 describes computer program setups for descriptive statistics; worthwhile reading even if you are not going to use the packaged program.

Weinberg, George H., and John A. Schumaker, *Statistics: An Intuitive Approach,* 3rd ed. Monterey, Calif.: Brooks/Cole, 1974. Chapter 8 is an excellent treatment of the "normal distribution."

HYPOTHESIS TESTING

5

In the previous chapters we have concentrated on basic statistical concepts, and mainly on how these concepts underlie our reasoning from data to descriptions of what the data are intended to represent. In the present chapter we turn to a discussion of a type of decision procedure that incorporates reasoning with statistics. In such cases the decision concerns a statement that we are attempting to assess by conducting a study. In short, this decision procedure concerns the role of statistics in hypothesis testing. Within this procedure we still use descriptive statistics and sampling statistics, but we use them in a rather special way.

KINDS OF HYPOTHESES

Modern science has taught us not to expect an absolute degree of certitude in statements that we make about the real world. Instead, we gauge certitude in terms of probability. In order to use probabilities for this purpose, we usually begin with statements that we wish to test in our studies; these statements are called *hypotheses.* In most psychological studies of communications or educational phenomena, however, two kinds of hypotheses are involved. One of these, called a *null hypothesis,* can be evaluated in terms of the probabilities that sampling statistics provide. The other, called a *research hypothesis,* is the actual research prediction that we want to test. The research hypothesis is always the logical opposite of the null hypothesis, such that if the null hypothesis is found to be relatively improbable (according to statistical criteria), then, by implication, the research hypothesis is taken as acceptable. We need these two kinds of hypotheses in order to bridge the gap between what sampling statistics can tell us about probabilities and the kinds of statements that we want to make about phenomena.

In order to make this discussion as simple as possible, let us depart from psychological measures and use an easily visualized example that relates samples to populations. Suppose that we have scooped up two handfuls of marbles and want to know if, in terms of the average weight of the marbles, the two handfuls have been drawn from the same population, or at least populations with equal means. These randomly drawn handfuls are our examples of *sam-*

ples. Whatever larger batch of marbles we drew them from we consider the *population*. Logically, if the handfuls were drawn randomly from the same population, we would expect the mean weight of marbles in each to be similar if not the same. Sampling error will be reflected in how much they differ if actually drawn from the same population. On the other hand, if their mean weights differ more than we might expect from sampling error, then we might decide that they came from different populations. In order to set ourselves up for this decision, we use the aforementioned null and research hypotheses.

The Null Hypothesis

When we make the statement that whatever differences are observed between sample means is due to sampling error, it is called a null hypothesis.

> 5.1 *Null hypothesis:* a statement that statistical differences or relationships have occurred for no reason other than laws of chance operating in an unrestricted manner.

In this particular example, we could state the null hypothesis in terms of the phenomenon under study—that the difference between the mean weights of the two samples of marbles is due to sampling error, and that they have the same population mean. Or else we can shorten the statement in terms of a mathematical expression,

$$\mu_l = \mu_r$$

where μ_l is the population mean represented in the sample of the marbles in the left hand, and μ_r is the population mean relating to the marbles in the right hand.

Recall in Chapter 4 how we saw that sampling statistics can provide us with estimates of the probability that certain sample characteristics could occur by chance (based on random sampling). In the present case this same logic applies in estimating probabilities of obtaining various differences *between* the means of the two samples. For example, we might estimate that in drawing successive pairs of samples we would expect to have a difference between the sample means as large as one-sixteenth ounce in 5 out of 10 cases of random sampling (*probability* $= .50$), whereas the likelihood of obtaining a difference as large as one-fourth ounce might only be 1 in 1000 cases ($p = .001$). Put into practical terms, we are able to calculate the probability that some observed difference in mean weights of the two samples could occur on the basis of random sampling. This puts us in a position to use probability as a basis for a decision about the acceptability of the null hypothesis. Sampling statistics aid us in gaining these estimates of probability.

The Research Hypothesis

Usually, however, our primary research interest is *not* in the prediction that some difference occurred by chance. Instead, we are interested in the acceptability of a statement that there is a difference, not just a chance one. In other words, our primary interest is in the acceptability of a statement that is the logical alternative to a null hypothesis; this statement is called the *research hypothesis*.

> 5.2 *Research hypothesis:* a statement expressing differences or relationships among phenomena, the acceptance or non-acceptance of which is based on resolving a logical alternation with a null hypothesis.

How, then, do we test the acceptability of a research hypothesis? Definition 5.2 points out the answer. The null hypothesis and the research hypothesis are taken as logical alternatives. We take the research hypothesis as acceptable (or supported) only if we have decided that the null hypothesis is unacceptable because of its low probability. Note that the emphasis here is on the null hypothesis being considered improbable. This is a key point, because our decision about the null hypothesis is the only link between what our statistical models tell us about probability, and the decision that we make concerning the research hypothesis.

Suppose, again, that we were studying the two handfuls of marbles. To put the problem in a practical context, let us presume that each handful represents a random sample of the day's output of a given marble factory and that we have good reason to predict that marbles produced by the two factories (I, II) do not have the same average weight. The average we want to consider is the mean of the population—not just the two handfuls. Our research hypothesis, then, might be stated as:

> There is a difference in average weight of marbles in the population produced by Factory I and the population produced by Factory II.

In mathematical form, the research hypothesis would be:[1]

$$\mu_I \neq \mu_{II}$$

The logical alternative to the above prediction would be stated in terms of the null hypothesis:

$$\mu_I = \mu_{II}$$

Sampling statistics would provide us with a basis for making a probability statement about the null hypothesis. If we found this cal-

---

[1] The sign $\neq$ is read "not equal to."

culation of probability to be rather low—say, .05 or less—we might take it as a basis for rejecting the null hypothesis. In other words, we would say that it seems unlikely that the difference between the two means could be due solely to sampling error. In rejecting this null statement, we would then accept its logical alternative, the research hypothesis.

A research hypothesis is either accepted or not on the basis of a decision about a null hypothesis. The decision about the null hypothesis is based on what sampling statistics tell us in terms of the probability with which the observed difference could have occurred under the terms of the null hypothesis, that is, due to sampling error. These steps will seem less cumbersome when seen in terms of the problem-method-results context of research designs.

THE PROBLEM

In terms of the research problem, a design involving hypothesis testing will incorporate a research hypothesis, a null hypothesis, and a probability level selected as a criterion for rejecting the null hypothesis.

Statement or Implication of a Research Hypothesis

The prediction of some difference or relationship may appear in many different kinds of statements, or it might not be directly stated at all in a research report. Typically, however, the statement of a research hypothesis represents a compromise between fulfilling two goals. On the one hand, the prediction is phrased in terms that will orient the reader to the problem under study; for example:

> Magazine readers and nonmagazine readers have different annual mean incomes.

On the other hand, the research hypothesis to be tested statistically must be an unambiguous mathematical statement such as,

$$\mu_r \neq \mu_n$$

A more detailed verbal statement might be something like:

> In terms of mean annual income, there is a difference in the mean incomes of two populations as reflected by a difference between the mean (M_r) of a sample of magazine readers and the mean (M_n) of a sample of nonmagazine readers.

The crux is that a research hypothesis, however stated, must imply a null hypothesis that is susceptible to a probability estimate.

Usually the need for precision in bridging the gap between the statement comprising a research hypothesis and what the researcher specifically means in terms of the study is served by presenting a series of definitions. In the above examples of an hypothesis, for example, precision is gained by definitions that answer such questions as: What exactly is meant by a *magazine reader* as distinguished from a *nonmagazine reader*? Are the *populations* restricted to some geographic area? What constitutes *annual income*? What is the time element involved, the period to which the statement applies?

Sometimes a researcher will use the statistical logic of hypothesis testing but does not claim beforehand to be testing specific predictions. Such cases sometimes involve a research problem stated in purpose form, for example:

> The major purpose of this study was to investigate variations in the behavior of a student answering a question when negative and positive feedback are provided as a teacher's evaluation of the answer.

In the above problem statement, research hypotheses are implied in the presumed relation between the feedback and the answers. It might be stated generally as:

> Verbal behavior under negative feedback $\neq$ Verbal behavior under positive feedback

Of course, the researcher would have to define the measures involved in the prediction, the exact conditions of feedback, and so on.

There may be times, too, when the statistical implication of a research hypothesis is somewhat secondary to the researcher's general problem. It is not unusual to find problem statements in the form of questions, for example:

> This research was designed to probe the general area of how people gain information about health problems. Specifically, answers were sought to the following questions:
> 1. Do people in the age category of 20–39 years differ from those in the 40–59 year range in the average degree to which they seek health related information?
> 2. Do Black families equated on a socioeconomic scale with White families differ in the average degree to which they seek health related information?
> 3. And so on.

The above study does not center on testing predictions, but when conducting statistical tests such as suggested in the questions of population differences, a research hypothesis of some greater-than-chance difference is implied, as well as a corresponding null hypothesis of chance difference.

Statement or Implication of a Null Hypothesis

In the previous example involving annual incomes, it was indicated that $\mu_r \neq \mu_n$, as a research hypothesis, implied $\mu_r = \mu_n$ as the null hypothesis. But suppose, for example, that the research hypothesis is

$$\mu_r > \mu_n$$

(the reader population mean is greater than the nonreader population mean)

The implied null hypothesis, then, is

$$\mu_r \leq \mu_n$$

(The reader population is equal to or less than the nonreader population mean)

In practice, the null hypothesis is frequently implied rather than stated. In fact, if there is nothing special about the null hypothesis — that is, it is clearly implied by the research hypothesis — a statement of the null explanation is apt to be redundant. The key feature of a null hypothesis is that it must be compatible with probability estimates provided by a statistical model, while at the same time a logical alternative to the research hypothesis.

Choice of a Probability Level

Bear in mind that we are interested in selecting a probability level as a criterion for rejecting a null hypothesis. What we are saying is that if the probability calculated for the null hypothesis is at this level — or even less probable — we shall reject the null hypothesis and accept the research hypothesis.

Sometimes a probability of .05 is taken as a level suitable for rejection of the null hypothesis. An actual value of probability is then calculated by statistical procedures and, if it is at or below this .05 level, it is taken as grounds for rejection of the null hypothesis. As a criterion for decision about the null hypothesis this level is often called the *rejection region,* or the *significance level.*

5.3 *Rejection region* (*significance level*): a level of probability set by the researcher as grounds for the rejection of the null hypothesis.

If a calculated value of probability is such that it falls within the rejection region, the researcher will interpret the difference or relationship as *statistically significant.*

5.4 *Significance:* the level of calculated probability was sufficiently low as to serve as grounds for rejection of the null hypothesis.

Often times the level set for significance is expressed in terms of a lower case letter p (meaning probability), a sign indicating "less than," and some particular probability value, for instance "$p < .05$," "$p < .01$." Thus, if a researcher reports that a difference was "significant at the $p < .05$ level," he or she is saying simply that the null hypothesis was rejected upon attaining this level, or a lower level, of calculated probability.

There is nothing sacred about the $p < .05$ level; it is set mainly out of convention in using statistical methods. There are, however, many cases where some other rejection region will be employed. Setting this rejection requires consideration of the types of error a researcher is willing to tolerate, a topic discussed subsequently.

THE METHOD

In terms of its statistical aspect, the method centers upon how we calculate values of probability associated with the null hypothesis. As mentioned earlier, this involves use of both descriptive and sampling statistics. For purposes of illustration, suppose that we were testing the hypothesis stated earlier:

Magazine readers and non-magazine readers have different annual mean incomes.

or:

$$\mu_r \neq \mu_n$$

We would use descriptive statistics to provide indexes of the characteristics of the samples of the two populations. Among other things, these indexes would include the two sample means, one for the readers (M_r), one for the nonreaders (M_n). We could call each of these means a *statistic* (in the sense of a sample characteristic), and we call their difference (that is, $M_r - M_n$) a statistic. What we want to know is the sampling error of the statistic, $M_r - M_n$.

Recall in Chapter 4 how we first considered the concept of the standard error of a mean (σ_M). This was a mathematical way of expressing sampling error. Without going into the details of calculation, consider the concept of the *standard error of the difference between means*. Just as a single sample mean can be thought of as a statistic, and its sampling error expressed in terms of the concept of standard error, so we can treat the difference between two sample means as a statistic and consider its standard error (that is, $\sigma_{\text{diff.}}$).

Figure 5.1 illustrates what a distribution of the sampling error of the mean difference might look like. Imagine that instead of just two samples, we had successively drawn and redrawn many pairs of samples. Each pair of samples has a mean difference (for example,

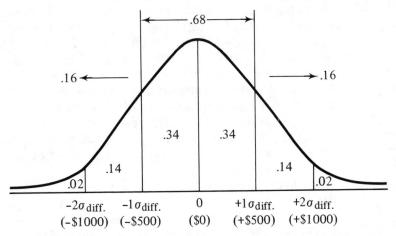

FIGURE 5.1 Sampling distribution of values of $M_r - M_n$ under the null hypothesis of $\mu_r = \mu_M$, and where $1\sigma_{\text{diff.}} = \$500$

$M_r - M_n$). If we took all of these values of mean differences, we could plot a distribution much like the one in Figure 5.1. Just as we saw how a distribution of sample means (Fig. 4.7) could have its own standard deviation, so can this distribution of mean differences. But in the present case, we call this standard deviation, a *standard error of the mean difference*. In reality, we do not draw successive pairs of samples, but calculate this value by use of a particular formula.[2]

Notice that the baseline of this distribution (Fig. 5.1) is divided into σ units. But this time these units represent standard errors of the mean difference ($\sigma_{\text{diff.}}$). We can also see how actual values of $M_r - M_n$ fit along this baseline because we would have calculated what $1\sigma_{\text{diff.}}$ is equal to in terms of actual measurement units. What this distribution represents is the different values of $M_r - M_n$ that we would expect because of sampling error, assuming that the curve is based on the null hypothesis, for instance, $\mu_r = \mu_n$ or $\mu_r - \mu_n = 0$. Put another way, if the two samples actually came from populations having the same mean, we could expect the kind of sampling error in values of $M_r - M_n$ as shown in Figure 5.1.

How, then, does probability enter the picture? Consider what we can interpret from the curve in Figure 5.1. Similar to the curves discussed in Chapter 4 (Fig. 4.4), its area represents probability. Thus we can pick particular segments of the baseline of this distribution and interpret them in terms of the probability that some value will fall within them. What, for example, would be the probability of obtaining a value of $M_r - M_n = \$500$ or greater? According to the sam-

[2] For example, $\sigma_{\text{diff.}} = \sqrt{\sigma_{M_r}^2 + \sigma_{M_n}^2}$

pling error expected under the null hypothesis, the probability would be approximately .32 as shown in Figure 5.1. That is, the value of $500 could be either in the tail of the curve *below* −$500, or −$1\sigma_{\text{diff.}}$, or in the tail *above* +$500, or +$1\sigma_{\text{diff.}}$. The probability in each of these tails is equal to approximately .16, so if we consider that either tail might apply, the combined probability is .32. What would be the probability of occurrence under the null hypothesis of obtaining $M_r - M_n = \$1000$, or greater? Again consulting the curve in Figure 5.1, we see that this value would fall in the tail below $-2\sigma_{\text{diff.}}$, or the tail above $+2\sigma_{\text{diff.}}$, and these combined areas would be interpreted as a probability of .04.

What we have seen, then, is the kind of a distribution that underlies how the probability of obtaining values of $M_r - M_n$ could be estimated. Again, this distribution represents the sampling error that would be expected under the null hypothesis. Hence, probabilities interpreted from it are the probabilities that we use when considering the hypothesis that $\mu_r - \mu_n = 0$, and that any $M_r - M_n$ differences we have observed are due solely to chance (sampling error).

In conducting actual analyses, we do not generally plot such curves. We calculate the value of $\sigma_{\text{diff.}}$, then along with a value of $M_r - M_n$, consult a table that gives us the probability value that would be expected under an appropriate curve. Chapter 6 on the t test will provide an example of an actual calculation.

THE RESULTS

The final step of the strategy for hypothesis testing involves interpretation of statistical results and, based on these results, stating a conclusion for the study. Ideally, this conclusion will be reasoned directly from the statistical results and will fulfill whatever goal was defined in the statement of the problem.

Interpreting Statistical Results

We have already seen in Figure 5.1 how, for a given difference between two sample means, we can estimate the probability of this occurrence under the terms of the null hypothesis. Let us now look again at this same distribution of sampling error under the null hypothesis, but this time consider how the rejection regions fit in.

In the preceding example of the distribution of the statistic $M_r - M_n$, we considered probability in terms of both tails of the curve. Since the research hypothesis ($\mu_r \neq \mu_n$) said nothing about the direction of the difference between the means of the two populations, we considered the likelihood that a particular observed value of $M_r - M_n$ would be on *either* side of the curve. Put into more practical terms, since our research hypothesis said simply that the two population means would be different, but did not state the direction of this difference, we might expect to find a value of $M_r - M_n$ in either tail of the curve. Suppose, then, that we had set $p < .05$ as the significance level; can we see this rejection region in terms of the distribution of M_r-M_n under the terms of the null hypothesis? The answer to this question is shown in Figure 5.2. Here we have the same curve as presented in Figure 5.1, but this time we are considering it only in terms of where the rejection regions fall.

Note first that the portion of the curve incorporating rejection areas totals .05, the level of probability that we have set for rejection of the null hypothesis. We have divided this .05 into areas of .025 in each tail because the hypothesis we are testing may involve a value of $M_r - M_n$ that could fall in either tail. What this curve tells us is, if $1\sigma_{\text{diff.}} = \$500$, a value of $M_r - M_n$ of either $-\$980$ or $+\$980$, or greater, will be within the region set for rejection of the null hypothesis. This is called a *two-tailed* test.

> 5.5 *Two-tailed test:* a nondirectional hypothesis test that incorporates rejection regions in both tails of the probability curve used for a given statistic.

Suppose, however, that the research hypothesis said something about direction—for example:

$$\mu_r > \mu_n$$

The null hypothesis is then:

$$\mu_r \leqslant \mu_n$$

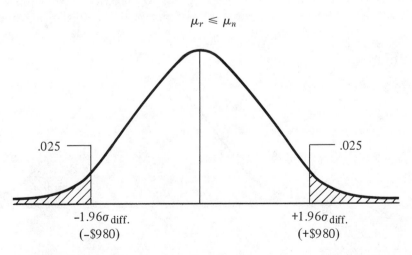

FIGURE 5.2 Rejection regions for a two-tailed test ($p < .05$) of the null hypothesis $\mu_r = \mu_n$ against the alternative hypothesis of $\mu_r \neq \mu_n$, and where $1\sigma_{\text{diff.}} = \$500$

This kind of a null hypothesis—that is, one incorporating direction—need not be tested against both tails of the probability curve. If we were to obtain a value of $M_r - M_n$ that would serve as a basis for rejecting this null hypothesis, then we would expect it to be only a positive value, that is, a difference in the direction predicted by the research hypothesis. Since in this case we are testing the null hypothesis only in terms of what might be found in one tail of the curve, the rejection region is located as shown in Figure 5.3. If this rejection region has been set at $p < .05$, then we locate this entire region only in the tail of the curve that would be applicable in considering rejection of the null hypothesis. In this case, according to the distribution shown in Figure 5.3, a value of $M_r - M_n$ equal to +$825, or greater (and no negative values) would serve as grounds for rejecting the null hypothesis. In practical terms, we are simply testing the null hypothesis against values of $M_r - M_n$ that could occur in one direction only. Those that might occur in an opposite direction could not be used as a basis for rejecting a null hypothesis. Obviously, then, in setting up a directional research hypothesis, the researcher must be very confident beforehand that there is no reason to expect differences in an opposite direction. This kind of an hypothesis test is called *one-tailed*.

>   5.6 *One-tailed test:* a directional hypothesis test that incorporates a rejection region in only one tail of the probability curve used for a given statistic.

Again, a one-tailed test is never used (or should never be used) unless there is very good reason to make a directional prediction. This reason is not especially a function of wanting to make a directional prediction, but is mainly based on the confidence that an outcome in

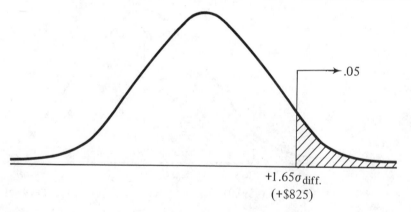

$+1.65\sigma_{\text{diff.}}$
(+$825)

FIGURE 5.3   Rejection region for a one-tailed test ($p < .05$) of the null hypothesis $\mu_r = \mu_n$ against the alternative hypothesis $\mu_r \neq \mu_m$ and where $1\sigma_{\text{diff.}} = \$500$

the opposite direction will not be obtained. If a one-tailed test is made in the absence of such confidence, the researcher is assuming the hazard of never really testing statistically an outcome that would be opposite to the predicted direction.

The statistical result of a hypothesis-testing study is a probability value determined for the null hypothesis. If this probability value falls within the region set for rejection, the researcher then has a basis for rejecting the null hypothesis and accepting, in turn, the research hypothesis. As will be seen in Chapter 6, we never really plot such curves as shown in Figure 5.2 and 5.3, but instead use tables for interpreting statistical results.

Stating the Conclusion

Given a decision regarding the null hypothesis, a final step involves reporting the outcome of the hypothesis test. Let us consider how some of the alternative outcomes might be stated.

Suppose that the calculated value of probability were greater than that set as the significance level (rejection region). For example, if the rejection region had been set at $p < .05$, but the statistical results indicated that $p > .05$, in terms of the null hypothesis, we might say:

It is likely that the two samples represent populations with equal means,
(or)
There is no reason to believe that the two samples represent different populations in terms of their means,
(or)
The observed difference between the two sample means was not statistically significant.

In terms of the research hypothesis, when the null hypothesis has not been rejected, we might write something such as:

There was no evidence to support the research hypothesis that. . .,
(or)
There was no reason to believe that. . . .

On the other hand, if the outcome were such that $p < .05$, the null hypothesis would be rejected. If we wished to say something about the null hypothesis, it might be

It is likely that the two samples represent populations with different means,
(or)
There is reason to believe that the two samples represent different populations in terms of their means,
(or)
The observed difference between the two sample means was statistically significant.

The conclusion for the research hypothesis might be phrased something like:

There was evidence to support the research hypothesis that. . .,
(or)
Results supported the research hypothesis that. . . .

ERRORS IN HYPOTHESIS TESTS

Thus far you should be able to see that a good share of the logic in hypothesis testing involves nothing more than setting up a decision procedure for evaluating a research hypothesis. In discussing this logic, we have put the stress on what circumstances lead to support of a research hypothesis. A vital part of this logic can also be seen in terms of the kinds of *error* that we attempt to guard against in this decision procedure.

Type I and Type II Error

The two main kinds of error considered in the decision procedure for hypothesis testing are

5.7 *Type I error:* rejecting a null hypothesis when it should have been the acceptable alternative.
5.8 *Type II error:* accepting a null hypothesis when it should have been the rejected alternative.

To aid us in gaining a conceptual understanding of these two types of error, let us return to the probability distribution of different values of $M_r - M_n$ expected under the null hypothesis. The curve in Figure 5.4 is what we would expect if the null hypothesis were true. In terms of this curve, what we have set aside as a rejection region of $p < .05$ can also be seen as the probability of committing Type I error. Why is this the case? For one thing, the distribution in Figure 5.4 represents what we would expect in terms of the null hypothesis. When considering Type I error, we assume for the moment that this curve is the true or acceptable alternative, as compared with the research hypothesis. What we are showing in the rejection region is the probability of rejecting this distribution as the acceptable alternative, or, the probability of Type I error. Put into simpler terms, whatever level of probability we set as a basis for rejecting the null hypothesis is also the probability of our committing a Type I error.

Type II error is a little more difficult to conceptualize. Note first, as indicated in Definition 5.8, that Type II error implies that some $M_r - M_n$ distribution other than that for the null hypothesis is the acceptable alternative. To conceptualize an example of this other distribution, let us assume that $\mu_r - \mu_n$ is really equal to $\$1500$. That is, the two population means are indeed $\$1500$ apart. We shall then

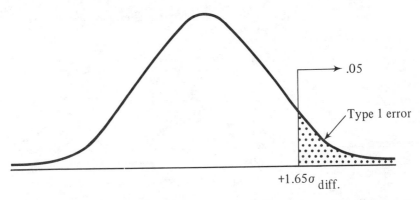

FIGURE 5.4   Rejection region for a one-tailed test ($p < .05$) of the null hypothesis, where the area (dotted) of the rejection region equals the probability of Type I error

compare these two distributions—that is, the null and the true alternative—on the same graph, with a common baseline, in Figure 5.5. Given these two distributions, suppose that we set the significance level at $p < .05$. The area under the null hypothesis distribution, but to the right of the significance level, is the probability of Type I error. By contrast, the area under the distribution of the true alternative, but to the left of the significance level, is the probability of Type II error.

If the curves in Figure 5.5 tend to confuse you, try to remember the following. Type I error is where the null hypothesis was really true (thus we consider area in the curve under the null hypothesis), but where we have rejected it (thus the area to the right of the significance level). By contrast, Type II error is where the alternative hypothesis was really true (thus we consider area under the curve of the true alternative), but where we have accepted the null hypothesis instead (thus the area to the left of the significance level).

With all other factors being equal, consider what the consequences are of changing the significance level. Suppose, for example, that we set it as $p < .01$. This would be shown by moving the significance level to the right in Figure 5.5, thus decreasing the probability of Type I error, but increasing the probability of Type II error. The converse is also true. Suppose that the significance level were set at $p < .10$. This would have the effect of moving the significance level to the left in Figure 5.5, thus increasing the probability of Type I error, but decreasing the probability of Type II error.

Power

There is an additional main factor that separately affects Type II error, however. This is *statistical power,* a probability that is also shown in Figure 5.5.

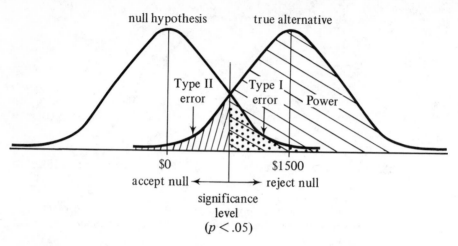

FIGURE 5.5 Comparison of $M_r - M_n$ distributions under the null hypothesis and under a true alternative

5.9 *Power:* the probability of rejecting a null hypothesis that is, in fact, false.

Since power is based on the idea of the null hypothesis being false, we are interested in some area within the distribution of the true alternative. Also, since power is based on our acceptance of this alternative distribution, then we are considering all of the area to the right of the significance level shown in Figure 5.5. One way of considering this area is that it is equal to 1−the probability of Type II error. In other words, all other factors being equal, if we increase the power of a test, we are in effect changing the picture shown in Figure 5.5 so that the amount of Type II error is accordingly reduced. All this would be independent of any effect on the probability of Type I error.

What we want, then, in a statistical test is power—the probability of rejecting a null hypothesis that is, in fact, not the acceptable alternative. If we were somehow to diminish power, we would be more prone to Type II error, that is, the probability of accepting the null hypothesis when it should be rejected. How do we get power? Suffice to say that we gain power from certain factors in the design of the experiment (such as sample size) and from the kind of statistical models that we use for our calculations for the distributions.

Given a situation where a particular power is attained, the setting of the significance level is then a compromise between Type I and Type II error. That is, if we increase the probability of Type I error, we decrease the probability of Type II error, and vice versa. Usually this decision is based on the kind of error that the researcher is willing to tolerate. Suppose that we were engaged in so-called pilot research. We are more or less trying out an experimental design to see if some gross effects are observed. Here we might set the significance level at .10 or .20, since we are not greatly worried about the consequences of rejecting a true null hypothesis. We are simply looking for grounds to conduct a further, more rigorous, experiment. On the other hand, suppose that we could not tolerate Type I error to any great degree. Perhaps the experiment tests a prediction that we want to be highly confident about when the null hypothesis is rejected. Thus we might set the significance level at .01 or even .001.

A NOTE ON FAILURE TO REJECT THE NULL HYPOTHESES

If a study results in failure to reject a null hypothesis, the researcher has not really "proved" a null hypothesis, but has failed to find support for the research hypothesis. It is not unusual to find studies with negative outcomes where the researcher has placed a great deal of stock in "acceptance" of null hypotheses. Such interpretations, strictly speaking, are in error because the logic of a research design incorporates the testing of some alternative (research hypothesis) against the status quo (null hypothesis). Although failure to find support for the alternative does leave one with the status quo, it *does not* rule out other possible alternatives. Put into practical terms, be skeptical of interpretations of unrejected null hypotheses.

SUMMARY

Hypotheses can concern predictions of relationships or differences. In the present chapter, the examples center upon predictions of differences. When concerned with such a prediction, we state a research hypothesis. Alternatively, for statistical purposes, we state a *null hypothesis,* which says that whatever difference we have observed between samples is due to sampling error. Sampling statistics provide us with a basis for estimating the probability that some observed difference between samples would be expected upon the terms of the null hypothesis—due to sampling error. If this obtained value of probability is equal to or less than some value preset as a *rejection region (significance level),* we reject the null hypothesis in favor of its logical alternative, the research hypothesis.

In stating a problem in hypothesis form, the research hypothesis must imply an unambiguous null hypothesis in terms of the population characteristics to be compared. Along with setting up these hypotheses, a significance level is also set. Sampling statistics then provide a basis for estimating the probability value that will eventually be compared with the significance level. Given this comparison, a decision is made with regard to the null hypothesis, and this in turn serves as a basis for a decision about the research hypothesis.

Much of the practical logic in hypothesis testing can be seen as a decision procedure. In considering this procedure, we often center upon the kinds and degrees of error that are involved. Two main types of error are *Type I,* the probability of rejecting a null hypothesis that should have been the acceptable alternative; and *Type II,* the probability of accepting a null hypothesis that should have been the rejected alternative. Finally, the *power* of a test is seen as the probability of rejecting a null hypothesis that is the unacceptable alternative.

SUPPLEMENTARY READINGS

Most basic statistics texts have introductory chapters on hypothesis testing; here are several:

Kerlinger, Fred N., *Foundations of Behavioral Research,* 2nd ed. New York: Holt, Rinehart and Winston, 1973. Chapter 2 introduces hypotheses as related to the research problem; Chapter 12 introduces the statistical nature of hypotheses.

Levin, Jack, *Elementary Statistics in Social Research,* 2nd ed. New York: Harper & Row, 1977. Chapter 8 introduces materials on differences between means.

McCall, Robert B., *Fundamental Statistics for Psychology,* 2nd ed. New York: Harcourt Brace Jovanovich, 1975. See Chapters 8, 9, and 10 on hypothesis testing.

Weinberg, George, and John A. Schumaker, *Statistics: An Intuitive Approach,* 3rd ed. Monterey Calif.: Brooks/Cole, 1974. Chapter 11 is comprehensive and clear.

DIFFERENCE ANALYSIS
PART 2

THE *t* TEST

6

In the previous chapter we saw how statistical logic is brought to bear in testing statements about population differences. Given a research hypothesis, for example, that two population means differ, this implies a null hypothesis of no difference. Under the terms of this null hypothesis we would expect that any differences between the means of samples of these two populations is due to sampling error. Using statistical reasoning, we can calculate the probability that a difference between sample means as large as what we have observed would be expected under the terms of the null hypothesis, that is, if due to sampling error. If we find this probability to be as low as, or lower than, a level of probability set as a criterion for rejecting the null hypothesis, we reject the null hypothesis and accept its logical alternative, the research hypothesis, and conclude that the sample means reflect different population means.

In this chapter and in three subsequent chapters we shall review in some detail the various statistical procedures used in tests of differences. These models each provide a basis for reasoning statistically from sample distributions in order to gain an estimate of the probability that whatever difference was observed between or among samples is due to sampling error.

The present chapter centers upon the *t* test, and how it is used for testing the difference between two population means, based on the observed difference between two sample means and their distributions. Although this is a common application of the *t* test, the reader should be aware that the *t* test itself has several other uses.

A SAMPLE STUDY

To illustrate how the *t* test of means fits into a research plan, we shall first describe a sample study. The data have been deliberately simulated in skeletal form for ease of understanding.

Problem

The experimenter is interested in assessing whether children's perception of aggressive behavior in a television cartoon will affect aggressive tendencies in their behavior. One theory reasons that children will imitate the aggression that they see in the cartoons, thus viewing will lead to an increase in children's tendencies toward aggressive behavior. A counter theory poses that any aggressive tendencies children might have are vicariously released by seeing the cartoon, hence behavior will be markedly less aggressive than usual after viewing such a cartoon. Considering both of these theories, the researcher predicts that viewing aggression in a cartoon will have some effect on children's aggressive tendencies, but there is no prediction whether this will be an effect of increase or decrease in these tendencies. In more specific terms, the research hypothesis is that the mean ratings of the amount of aggressive tendencies of children who have seen an aggressive cartoon (Group I) will be different from the mean ratings of children who have not seen such a cartoon (Group II); that is:

$$\text{R.H.:}\ \mu_I \neq \mu_{II}$$

The implied null hypothesis is

$$\mu_I = \mu_{II}$$

The researcher sets $p < .05$ (two-tailed) as the significance level.

Method

From a kindergarten class, ten children are randomly selected to serve as subjects (Ss) in an experimental design.[1] Of these, five are randomly selected to serve in a group that sees a 10-minute televised cartoon in which the central character's behavior is predominately aggressive. The remaining five children see an alternative cartoon under exactly the same conditions, except that there is no aggressive behavior depicted.

Following exposure to the cartoons, both groups of Ss are placed in a playroom where their behavior is observed through a one-way mirror and rated by several judges. The ratings are in a numerical form for which the experimenter can assume the power of an interval scale, and the distributions presumably fit the normal curve model. Each child is assigned a value that represents the mean of the judges' ratings of aggressiveness. Given the measures of the two samples, the means are calculated:

$$M_I = 57 \qquad M_{II} = 52$$

The sampling statistic t will be used to test the null hypothesis against the alternative research hypothesis. A value of t is calculated according to an appropriate formula (as in Table 6.1). Values of t

[1] Again, the size of sample is intentionally small for purposes of illustration.

themselves have a sampling distribution that is used as a basis for estimating the probability that a particular value of *t* would be expected under the terms of the null hypothesis. In other words, under the null hypothesis, what is the probability that the value of *t* could occur because of sampling error? In this particular case (calculations in Table 6.1), $t = 2.50$. The probability of obtaining this value as a function of sampling error is $.02 < p < .05$, according to an appropriate table of the sampling distribution of *t* (Table 6.2)[2] A practical way to think about sampling error here is that if this comparison were repeated 100 times, we would expect that a *t* of 2.50 would occur between 2 and 5 times of these 100 times as a function of sampling error.

Results

Since the probability value meets the criterion for significance (that is, $p < .05$), the null hypothesis is rejected in favor of the research hypothesis. That is, $\mu_{I} \neq \mu_{II}$ is taken as the acceptable alternative. With this support, the researcher concludes that the different cartoons have caused the two groups of children, who were formerly sampled from the same population, now to represent different populations in terms of the measurement taken. The main conclusion is stated as:

> There is evidence to support the prediction that children viewing aggressive behavior in a televised cartoon will have a different tendency to engage in aggressive play than children who see a nonaggressive cartoon.

THE LOGIC UNDERLYING *t*

Given this sample experiment, just what has been involved in the process of reasoning with the *t* test? We can answer this question from two standpoints—what goes into the calculation of *t*, and how values of *t* are interpreted.

Calculation of *t*

The logic in calculating *t* is similar to our discussion of the hypothesis test of difference presented in Chapter 5. As a test of the significance of difference between means, the general formula for *t* is given as follows:

$$t = \frac{M_{I} - M_{II}}{\sigma_{\text{diff.}}}$$

where M_{I} and M_{II} are the sample means and $\sigma_{\text{diff.}}$ is the standard error of the difference between the means.

---

[2] $.02 < p < .05$ is read as: "probability greater than .02 but less than .05."

In the specific formula for calculating t, the numerator $(M_I - M_{II})$ is almost always the same.[3] It is the calculation of the denominator $(\sigma_{\text{diff.}})$ that varies in different formulas.

The concept of $\sigma_{\text{diff.}}$ is the same as discussed in Chapter 5. Again, imagine a distribution formed by drawing successive pairs of samples and plotting the $M_I - M_{II}$ value in each case. The standard deviation of this distribution is called the standard error of the mean difference, or simply: $\sigma_{\text{diff.}}$ For the t test, we calculate $\sigma_{\text{diff.}}$ in somewhat different ways, depending on the sizes of the samples, whether the variances of the two samples are approximately the same, and whether the two sets of measures are somehow related.[4] Table 6.1 shows how t would be calculated for the sample experiment. Note in the denomi-

Table 6.1   **Example of the Calculation of t (Skeletal Data)**

Grouped Data

I	II
61	50
53	48
59	56
55	54
57	52

M 57 52
n 5   5

Deviations from Sample Means

d_I	d_{II}
4	-2
-4	-4
2	4
-2	2
0	0

Deviations Squared

d_I^2	d_{II}^2
16	4
16	16
4	16
4	4
0	0
Σd^2 40	40

General Formula

$$t = \frac{M_I - M_{II}}{\sigma_{\text{diff.}}}$$

Specific Formula

$$t = \frac{M_I - M_{II}}{\sqrt{\left(\dfrac{\Sigma d_I^2 + \Sigma d_{II}^2}{n_I + n_{II} - 2}\right)\left(\dfrac{n_I + n_{II}}{n_I \cdot n_{II}}\right)}}$$

$$= \frac{5}{\sqrt{\left(\dfrac{40 + 40}{5 + 5 - 2}\right)\left(\dfrac{5 + 5}{5 \cdot 5}\right)}}$$

$$t = \frac{5}{2} = 2.50$$

With $d.f. = 8$, $.02 < p < .05$

[3] Strictly speaking, the numerator reads "$M_I - M_{II}$" if the null hypothesis is $\mu_I = \mu_{II}$, otherwise the numerator reads: "$(M_I - M_{II}) - (\mu_I - \mu_{II})$".
[4] "Related" refers to *correlation* between two sets of measures, a topic treated in Chapter 10. Correlated measures would be involved if the two sets of measures were on the same Ss, or else the two sets of Ss were somehow matched on traits relevant to the measure.

nator of the formula that standard error of the mean difference ($\sigma_{\text{diff.}}$) is estimated from the variance observed in the two samples. The two main ingredients in this case are the sum of squared deviations about the two means (that is, Σd_2) and the sample sizes (*n*). All formulas for the calculation of *t* accommodate sample variability in one way or another.[5] Just as our calculation of standard error of the mean (σ_M) is affected by sample variance, so is our calculation of the standard error of the difference between means ($\sigma_{\text{diff.}}$).

Interpretation of *t*

In Chapter 5 the difference between means and the $\sigma_{\text{diff.}}$ was interpreted in terms of the normal probability curve. This same curve can be used for interpreting *t* values when the sample sizes are relatively large, say, 50 to 60 cases each. But as sample sizes grow smaller, the sampling distribution of *t* departs from normality; its center becomes increasingly peaked, and its tails thicker. These changes in shape have consequences when determining values of probability for different values of *t*. Rather than try to depict all different shapes of the sampling distribution, it is more convenient to present it in tabular form. When interpreting a particular value of *t* for given sample sizes, the researcher consults a tabular form of a *t* distribution, such as shown in Table 6.2.

Notice that to enter a table of *t* (Table 6.2), it is necessary to consider what are called *degrees of freedom (d.f.)*. In this table, degrees of freedom reflect sample sizes.[6] When two independent sam-

Table 6.2   **Example of a Table of *t* (See pp. 195–196 for a more standard, complete, version)**

d.f.	Probability				
	.20	.10	.05	.02	.01
1	3.08	6.31	12.71	31.82	63.66
2	1.89	2.92	4.30	6.97	9.93
. . .	. . .	. . .	. . .	. . .	. . .
8	1.40	1.86	2.31	2.90	3.36
. . .	. . .	. . .	. . .	. . .	. . .
18	1.33	1.73	2.10	2.55	2.88
. . .	. . .	. . .	. . .	. . .	. . .
Infinity	1.28	1.65	1.96	2.33	2.58

[5] For relatively large independent samples the estimate of $\sigma_{\text{diff.}}$ is equal to $\sqrt{\sigma_{M_\text{I}}^2 + \sigma_{M_\text{II}}^2}$.
[6] Degrees of freedom enter into many different statistical procedures, as will be seen in subsequent chapters. The basic concept is not far removed from what the name implies. For example, given an array of scores of the size *N* and with a given mean, all scores except one could be varied at will and still preserve the same mean, so long as the one score compensated for the others in order to yield that mean. In this case *d.f.* is equal to $N - 1$. Given two subgroups of the combined size *N*, the means of the two subgroups could be preserved while varying all save one score in each subgroup; hence, *d.f.* here is equal to $N - 2$, as in the current example.

ples are being considered (as in the example experiment), degrees of freedom are equal to the sūm of the two sample sizes, minus the value 2 (that is, $n_I + n_{II} - 2$).[7] The different rows in the t table, each showing a different value of *d.f.*, accommodate the changes in the sampling distribution of t as the sizes of samples may vary. Given the row corresponding to the *d.f.*, for a particular comparison, the researcher then finds where the calculated value of t falls in terms of probability values. In the example experiment, for *d.f.* = 8, a t of 2.50 falls between the .05 and .02 columns. This means that probability will be reported as $.02 < p < .05$. In this same row, a value of t exceeding 3.36 would be reported as $p < .01$. But if t were less than 1.40, probability would be designated as $p > .20$.

While Table 6.2 is labeled for two-tailed tests, it can be adapted to one-tailed tests by halving the probability values at the top of each column. Thus if our t were equal to 2.50 and we were using a one-tailed test with 8 *d.f.*, probability would be $.01 < p < .025$. (We are still between the columns labeled .02 and .05, but we halve these probability values to .01 and .025, respectively.)

By examining Table 6.2 further, we can see the consequences that varying sample sizes will have on the interpretation of t. Note, for example, that with 8 *d.f.*, t must be 2.31 or greater to be associated with $p < .05$. But if the samples were such that *d.f.* = 2, then a larger value of t (4.30) would be necessary to meet the $p < .05$ criterion. When considering solely the interpretation of the sampling distribution of t, as sample sizes get smaller and all other factors are equal, a larger value of t is necessary to gain a particular value of probability. On the other hand, as sample sizes get larger, a smaller value of t would suffice. But as the sampling distribution of t approaches normality, further increases in sample size no longer have this particular effect. When sample sizes total roughly 100 or greater, many researchers use the normal curve model for interpreting t. This would be the same as entering Table 6.2 for *d.f.* equal to infinity, which is the same as the normal curve model (Chapter 4).

The sampling distribution of t tells us the probability that, in accord with the null hypothesis, we could expect a particular value of t. Put another way, the sampling distribution tells us what sampling error to expect in values of t.

SOME GENERAL CONSIDERATIONS OF t

Given what we have said about the t test, what are some characteristics of t that are useful to remember? For one thing, as logically

[7] For correlated measures, *d.f.* is equal to the number of *pairs* of scores minus the value 1.

suggested by the research hypothesis, as the difference between the sample means $(M_1 - M_{11})$ is increased, and all other factors are equal, the value of *t* will increase. And as indicated by the sampling distribution of *t*, as *t* grows larger, associated values of probability (*p*) grow smaller. As can be seen in the calculation (Table 6.1) of the denominator of the *t* formula, the greater is the variation among the scores of each sample and, all other factors being equal, the greater will be the estimate of $\sigma_{\text{diff.}}$, and hence the smaller the value of *t*. In other words, the more variability there is among scores in a sample, the more difficult it will be to gain a *t* for a particular significance level.

One logical question sometimes comes to mind when people think of likely extensions of studies using *t* tests: What if we want to compare more than two groups on a dependent variable? The answer is that we employ a technique called *analysis of variance,* the topic of the next two chapters.

Finally, along another line, what if we want to compare two groups in terms of more than one dependent variable? There are several alternatives, the most typical of which is *multiple discriminant analysis,* the topic of Chapter 15.

SUMMARY

The *t* test is a statistical model that can be used for testing the significance of difference between the means of two populations, based on the means and distributions of two samples. The logic of *t* is similar to the test of the difference between population means as discussed in Chapter 5; that is, it is a ratio between the sample mean difference and the standard error of that difference. The unique features of *t* are found in the various methods for estimating the standard error of the mean difference and in the fact that the sampling distribution of *t* departs from normality when small samples are employed. Given a calculated value of *t*, this value is interpreted for its probability of occurrence in testing a null hypothesis against an alternative research hypothesis. If this probability value is equal to or less than the level set for significance, the null hypothesis is rejected in favor of the research hypothesis.

SUPPLEMENTARY READINGS

Ferguson, George A., *Statistical Analysis in Psychology and Education,* 4th ed. McGraw-Hill, 1976. Chapter 12 includes *t*-tests.

Nie, Norman H., and others, *Statistical Package for the Social Sciences,* 2nd ed. New York: McGraw-Hill, 1975. Chapter 17 describes computer program setups for *t*-tests.

Weinberg, George H., and Schumaker, John A., *Statistics: An Intuitive Approach,* 3rd ed. Monterey, Calif.: Brooks/Cole, 1974. Chapter 14 is an excellent introduction to the *t*-test.

SINGLE-FACTOR ANALYSIS
OF VARIANCE

7

Whereas the t *test is usually used in testing the difference between two population means based on differences found between sample means, analysis of variance is used when the research hypothesis incorporates two or more population means and tests of differences among their respective sample means.*

SINGLE-FACTOR MODEL

What does single factor mean? Consider first that the term factor can mean the same thing as independent variable. In our examples in previous chapters we have seen an independent variable as involving a distinction in treatments administered by the researcher to the different groups of subjects involved in an experiment. Thus, in the preceding chapter, the kind of cartoon that the children saw was the independent variable. We would describe this independent variable as having two levels: viewing a cartoon showing no aggression and viewing a cartoon showing aggression. In that example the research was concerned only with the two post exposure sample means and whether or not they reflected the same or different population means. Suppose, however, that the experiment had three levels of the cartoon factor. This would involve a statistical model that could accommodate a test of more than two means. As described in the subsequent example, single-factor analysis of variance would be the likely statistical model in this case.

A SAMPLE STUDY

For the present purposes we shall extend the hypothetical experiment described in the preceding chapter. Again, the samples are intentionally small and simulated for ease of presentation.

Problem

Given the finding that viewing aggressive actions in a televised cartoon will affect aggressive tendencies in a child's own behavior, a further question is raised. Will there be such an effect if the aggressive behavior of the cartoon character is punished as a part of the cartoon drama? Seemingly, if children see that aggression leads to punishment then perhaps they will avoid such a tendency in their own behavior, or at least lessen it.

A second experiment is designed to study this problem. This time, three groups of children are incorporated into the study. Again, the children's behavior will be assessed following exposure to a cartoon. But this time the groups will vary according to whether a cartoon portrays no aggression, aggression that goes unpunished, or aggression that is punished. The prediction is that after seeing the different cartoons, the groups will have different mean tendencies to engage in aggressive play.

In mathematical form, the research hypothesis is the overall statement:

$$\mu_I \neq \mu_{II} \neq \mu_{III}$$

The implied null hypothesis is the overall statement:

$$\mu_I = \mu_{II} = \mu_{III}$$

The level set for rejection of the null hypothesis is $p < .05$.

Method

From a kindergarten class, 15 children are randomly selected to serve as Ss in the experiment. In turn, they are randomly assigned to the following groups, each with a respective treatment.

Group I ($n = 5$) sees a 10-minute televised cartoon in which the central character shows no aggressive behavior.

Group II ($n = 5$) sees a 10-minute televised cartoon in which the central character engages in aggressive behavior, and the cartoon does not include any punishment of this behavior.

Group III ($n = 5$) sees a 10-minute televised cartoon in which the central character engages in aggressive behavior, but this behavior is subsequently punished.

The independent variable, then, comprises the differences in the above three treatments, or three levels of the cartoon factor. The dependent variable is the Ss' tendency to engage in aggressive play. As in the previous example (Chapter 6), the measure is a numerical rating based on observations by judges.

Means for the three samples are

$$M_I = 51 \qquad M_{II} = 58 \qquad M_{III} = 53$$

The researcher uses analysis of variance as a method for making a probability statement about the null hypothesis. The calculations yield a statistical value called F. This value can be interpreted in a sampling distribution to determine its probability associated with occurrence under the null hypothesis. In this study (calculations in Table 7.1), F is equal to 16.25. The associated probability value (Table 7.2) is $p < .01$. That is, under the hypothesis of sampling error (null hypothesis), we would expect to find the present differences among sample means in fewer than 1 out of 100 cases of random sampling.

Results

Since the obtained probability value meets the criterion for statistical significance, the null hypothesis is rejected in favor of the research hypothesis. The researcher concludes:

> There is evidence that a child's tendency to engage in aggressive behavior will be differentially affected under conditions where a televised cartoon portrays no aggression, unpunished aggression, or punished aggression.

THE LOGIC UNDERLYING F

A first point of special note is that the hypothesis to be tested is considered as an overall statement. That is, analysis of variance will tell us only if there is significant variation among the means in that total statement; it will not tell us about the comparison of individual means, that is, μ_I versus μ_{III}, and so on. In other words, analysis of variance centers upon the question of whether the three samples represent the same population in terms of their means.

The Calculation of F

In the analysis of the sample experiment we could see how the mean scores varied for each of the three groups. Logically, the more that these groups would differ from each other in terms of the trait being measured, the more variability we would expect among their means. We can depict such variability statistically in terms of how the group means vary about a *grand* mean, that is, a mean of all of the groups put together. In fact, we can calculate the groups' variance about the grand mean. In analysis of variance this is called the *between-groups* variance. More specifically, the more difference that there is among the group means, the greater would be the value of between-groups variance. What we need next is some way to interpret this amount of variance—namely, what is the probability of it occurring under the terms of the null hypothesis (that is, due to sampling error)?

Suppose that we assumed that all three groups came from the same population. If this were so, how much variability among individual scores would we expect in this population? In analysis of variance, this variability is actually estimated. It is based on the variability that we observed within each of the groups—aptly called *within-groups* variance.

The key point in analysis of variance is that if there are no differences among the groups, then the between-groups variance and the within-groups variance will be approximately equal. To put this another way: The more that a value of between-groups variance exceeds the within-groups variance, the greater is the probability that the groups represent different populations.

With the foregoing in mind, consider the definition of F:

$$F = \frac{\text{variance between groups}}{\text{variance within groups}}$$

If the null hypothesis were true and there were no sampling error whatsoever, we would expect F to equal 1.0. But in reality we face the prospect of sampling error; therefore we again use a sampling distribution to gain a probability statement. Under the assumption of the null hypothesis, the sampling distribution of F tells us the probability that we can expect the between-groups variance to be so many times greater than the within-groups variance. As with other hypothesis-testing statistics, if this probability meets a preset criterion level, we shall take it as grounds for rejection of the null hypothesis.

Table 7.1 illustrates how F could be calculated for the sample experiment. Notice that most of the calculations refer to sums of squares (SS). The total SS reflects squared deviations of all scores about the grand mean, without any regard to group assignment. This total amount of squared deviation is partitioned into two components: squared deviations between the groups and the grand mean, and squared deviations of the scores within groups about their own means. These component squared deviations are reduced to mean squares (Ms), which are actual calculations of variance. The degrees of freedom used to make these reductions represent the number of groups minus one (between $d.f.$), and the number of Ss minus one in each group times the number of groups (within $d.f.$).

Usually the components of variation used to calculate F are displayed in a summary table similar to the one shown in Table 7.1.

Interpretation of F

As mentioned above, we use a sampling distribution of F to determine the probability that a particular value of F could occur under the terms of the null hypothesis. This sampling distribution changes quite radically according to the degrees of freedom in the numerator and denominator of the F ratio (in this case, reflecting the number of

Table 7.1   **Example of a Single-Factor Analysis of Variance (Skeletal Data)**

Total SS (Sum of Squares)			General Formula

Score deviations (d_g) from grand mean (54)

$$F = \frac{MS \text{ between}}{MS \text{ within}}$$

Group I	d_g	$d_g{}^2$
49	−5	25
52	−2	4
52	−2	4
53	−1	1
49	−5	25

Summary Table

Source	SS	d.f.	MS	F
Between	130	2	65	16.25
Within	48	12	4	
Total	178			

Group II		
56	2	4
57	3	9
57	3	9
60	6	36
60	6	36

Within SS

Score deviations (d_w) from group means:

I	I	III
−2	−2	1
1	−1	−1
1	−1	3
2	2	−3
−2	2	0

Group III		
54	0	0
52	−2	4
56	2	4
50	−4	16
53	−1	1

Deviations squared $(d_w{}^2)$:

4	4	1
1	1	1
1	1	9
4	4	9
4	4	0

Total SS = 178 = $(\Sigma\ d_g{}^2)$

Sums	14	14	20

Within SS = 14 + 14 + 20 = 48 $(\Sigma\ d_w{}^2)$

Between SS

		I	II	III
Group means...	(M)	51	58	53
Group deviation from grand mean...............	(d_b)	−3	4	−1
Squared deviation....................................	$(d_b{}^2)$	9	16	1
Group n times $d_b{}^2$....................................	$(nd_b{}^2)$	45	80	5

Between SS = 45 + 80 + 5 = 130 $(\Sigma\ nd_b{}^2)$

groups being compared, and the size of the groups). An abbreviated version of the F sampling distribution is presented in Table 7.2. This table is entered at the intersection of a column and row representing the degrees of freedom for the numerator and denominator of the F ratio, respectively. In this particular example of an F table, the numbers in boldface type represent the size of F necessary for a probability of .01.

For the F of 16.25 in the sample experiment and with $d.f. = 2/12.$,

Table 7.2   **Example of a Table of F** (*See pp. 197–198 for a complete version*)

d.f. Denominator	d.f. Numerator				
	1	2	. . .	8	. . . 12
6	5.99	5.14	. . .	4.15	. . . 4.00
	13.70	**10.90**	. . .	**8.10**	. . . **7.72**
. . .	. . .	. . .	. . .	. . .	. . . . . .
12	4.75	3.89	. . .	2.85	. . . 2.69
	9.33	**6.93**	. . .	**4.50**	. . . **4.16**
. . .	. . .	. . .	. . .	. . .	. . . . . .
16	4.49	3.63	. . .	2.59	. . . 2.42
	8.53	**6.23**	. . .	**3.89**	. . . **3.55**
. . .	. . .	. . .	. . .	. . .	. . . . . .
120	3.92	3.07	. . .	2.02	. . . 1.83
	6.85	**4.79**	. . .	**2.66**	. . . **2.34**
. . .	. . .	. . .	. . .	. . .	. . . . . .
Infinity	3.84	3.00	. . .	1.94	. . . 1.75
	6.63	**4.61**	. . .	**2.51**	. . . **2.18**

the probability value is less than .01. With $d.f. = 2/12$, an F of 3.89 would yield a probability of .05. We are, therefore, quite well below the level set for significance.

In considering the sample Table of F (Table 7.2), notice that as $d.f.$ would increase in the denominator of the F ratio, that a smaller F is necessary to gain significance at a particular probability level. What this reflects is what one gains by having larger samples in the various experimental groups. Notice, however, that such an increase in $d.f.$ soon pays diminishing returns; for example, with $d.f.$ of 1/120 an F of 3.92 is required for significance at the .05 level, but this required value of F only decreases to 3.84 with an infinite number of degrees of freedom in the denominator.

SOME GENERAL CONSIDERATIONS OF F

One useful point to remember about an F test is that the denominator represents our estimate of the variability among individuals in the population that we are studying. The numerator, by contrast, reflects whatever differences in variability are observed as a function of differences among the groups being compared. The ratio itself simply tells us by how many times the between-groups variability exceeds our estimate of the population variability. Finally, the sampling distribution of F gives us an estimate of the probability that the between-groups variance could be some multiple of the population variance. In other words, the F statistic is our basis for obtaining a probability statement referring to the null hypothesis. It is logical, then, that for a particular set of *d.f.* values, the larger is the value of F, the smaller is the value of p. That is, in terms of the null hypothesis, the larger the value of F, the less we would expect the null hypothesis to obtain. In two-group designs, F and t are related ($F = t^2$); they are assessing the same thing.

THE PROBLEM OF COMPARING INDIVIDUAL MEANS

When considering the results of the sample study, all that was said reflected only significant variation among the three groups. The analysis of variance results said nothing about μ_I being different from μ_{II}, or μ_{II} being different from μ_{III}, and so on. We only knew that somewhere among the means there was a significant difference (or differences) leading to the significant F. By rejecting the null hypothesis we are only taking the position that the overall statement of "$\mu_I = \mu_{II} = \mu_{III}$" is not acceptable.

Certainly, it is usually of research interest to know exactly whether each of these means is different from each other, and there are statistical procedures for making such comparisons. Often such methods are called *follow-up* or *post hoc* tests. These labels reflect a critical question in making such tests: In conducting individual comparisons of the means, are we reasoning according to predictions made in designing the experiment, or are we simply looking after the fact to see if any significant differences might show up? This is an important question because, if we have planned all along to test particular hypotheses of individual mean differences, we should bring the most powerful statistical tests to bear in assessing these differences. We must take great care when we are making tests after the fact, that is, when we are letting the data suggest hypothesis tests. Most statisticians agree that if we are simply looking for any differences that might result, we should be cautious in what differences we call significant. Being cautious involves the use of more conservative tests, often ones with *protection levels* that tend to lessen Type I error (that is, rejection of a null hypothesis that probably should not be rejected.

In practical terms, what the foregoing suggests is that we should take special care in selecting the particular type of test to use in making comparisons among individual means. There are many procedures available[1] for such tests, but it often becomes a relatively subjective decision to select the appropriate procedure to use. De-

---

[1] Some of the more frequently encountered ones are the *Turkey, Scheffé, Duncan,* and *Newman-Keuls* procedures. See the supplementary readings at the end of this chapter.

pending on how conservative a particular procedure is, a difference between means that might be significant with one approach might not be significant with another. Put into more practical terms, the conclusions that a researcher draws concerning individual mean comparisons are only as valid as the decisions that were involved in planning and selecting such tests—whether or not they were planned beforehand, and the amount of protection deemed realistic in the tests themselves.

Despite the problems of individual mean comparisons, if a researcher chooses to interpret any differences whatsoever among individual means, the statistical basis for such interpretations is incomplete without application of the proper follow-up tests.

To illustrate how individual mean comparisons facilitate describing the outcome of a study, suppose that we considered further the results of the sample experiment. The overall F ratio provided a basis for deciding that there was a significant difference among the three treatment means ($M_I = 51$, no aggression; $M_{II} = 58$, unpunished aggression; $M_{III} = 53$, punished aggression). After finding this difference, we are interested in individual comparisons among the means. Which ones are significantly different from one another? Consider that a follow-up test offering a moderate amount of protection against Type I error is applied (calculations are not shown).[2] The rejection level for these tests is set at $p < .05$. One way of reporting the results is as follows:

$$51_a^* \qquad 58_b \qquad 53_a$$

*(Means with common subscripts are not significantly ($p < .05$) different from one another.)

Here the follow-up tests indicate no difference between the group (I) who saw no aggression and the group who saw punished aggression (III), but both of these groups have means significantly different from the group who saw unpunished aggression (II). Notice that the researcher's concluding statement can now say much more than was reported earlier.

There is evidence that a child's tendency to engage in aggressive behavior will be differentially affected under conditions where a televised cartoon portrays unpunished aggression, as compared with no aggression or punished aggression. A *post hoc* analysis of the present results indicated that the effect of no aggression and punished aggression were not significantly different from one another, but that both were different from the condition where aggression went unpunished in the cartoon.

[2] D. B. Duncan, "Multiple Range and Multiple F Tests," *Biometrics*, XI (1955), 1–42.

SOME FUTHER NOTES ON SINGLE-FACTOR MODELS

Although what we have discussed thus far should provide you with an idea of the general logic involved in a single-factor analysis of variance, this has been a cursory picture at best. Within the scope of the present treatment, we can do no more than provide a few additional notes describing what some of the further considerations are when dealing with single-factor models. Many of these considerations also carry implications regarding the specific procedures for calculating and interpreting F ratios.

Random as Against Fixed Factors

One key distinction made when considering the different levels of a factor (independent variable) is whether these levels represent discrete variables or whether they are points along some type of continuous scale. For example, the different cartoons employed in the sample study represent discrete phenomena; as such, this variable is called a *fixed* factor. On the other hand, suppose that we had a series of cartoons that on a predetermined basis, was taken as representing different degrees of aggressiveness. In this case, the levels of the factor might be taken as sampling points along a continuum of aggressiveness in cartoons. We would call this a *random* factor; its levels are taken as a basis for "sampling" from a continuum of cartoon aggressiveness as an independent variable.

Levels of a Factor

Obviously, one way of increasing the complexity of a single-factor model is to incorporate additional levels of the factor into the design. The main criterion for defining levels is primarily a compromise between theory and the requirements of design. Suppose, for example, that we were interested in the length of the cartoon as the key factor in stimulating aggressiveness. As a random variable, we might want levels corresponding to 5, 10, 15, 20, and 30 minutes of cartoon length. So long as the different conditions all represent different levels of the same independent variable, the model remains a single-factor design.

Repeated Measures

In the present sample experiment, different groups of Ss were tested in each of the three levels of the cartoon factor. It is sometimes the case, however, that the same group of Ss is tested across the different levels of a factor. This again is another consideration that takes into account theoretical as well as design problems. In cases such as the sample experiment, it is advisable that an S be exposed to only one level of a factor, lest his or her exposure to one level influence exposure to another level. By contrast, consider a study where the independent variable is the length of time following exposure to aggression in a cartoon, and the dependent variable is again the S's tendency to engage in aggressive play. In this situation it is quite realistic to take successive measures on the same Ss; they might be taken, for example, at intervals of 30 minutes, 1 hour, 2 hours, and 4 hours. Such a design would involve *repeated* measures on the same subjects.

Analysis of Covariance

In most experiments we take all possible precautions to ensure that prior to whatever experimental treatments are administered, we have good reason to believe that the Ss in different groups are equal in terms of the behavior that is to be assessed. As in the sample experiment, such equalization is usually provided by carefully adhering to random assignment. But what happens if we cannot be sure of such assignment? Consider, for example, that in studying kindergarten children we had no other choice than defining three groups as simply three different schoolroom classes. Perhaps the original assignment of the children to these classes was on a more-or-less random basis, but we cannot be sure that the three classes are equal in terms of the behavior we want to study. Sometimes it is possible to equalize such pretreatment differences by a statistical method. This method is called *analysis of covariance.*

Suppose that in the sample experiment, the researcher was faced with using three groups that were in fact different kindergarten classes. And perhaps one of these groups has slightly more aggressive children than the others; thus there are group differences even before the experiment is conducted. The practical consequence of this situation is that if the experimenter finds a difference after the treatments, can it be attributed to the cartoons that were shown, or were the groups simply different to begin with? Use of analysis of covariance in this situation would proceed generally as follows.

Prior to exposing the children to the experimental cartoons, their aggressive tendencies are assessed by the judges. With such measures it is possible to have some idea of pretreatment intergroup differences among the children in terms of the trait to be studied. We call these pretreatment ratings a *control* variable (or a *covariate*).

The experiment is conducted as previously described. The groups are exposed to different treatments and subsequently are rated in terms of aggressiveness in behavior. We consider the posttreatment mean scores in this case as the *criterion* variable.

What analysis of covariance provides is a method by which we can remove pretreatment variations (as measured by the control variable) from the posttreatment means (criterion variable) prior to testing the significance of the posttreatment differences among the groups. In more simple terms, analysis of covariance provides a basis for ruling out pretreatment differences when our interest is in testing posttreatment differences. The significance test for analysis of covariance uses an F ratio, and this is interpreted for probability in a manner similar to a straightforward analysis of variance.

Obviously, the control variable is of little use unless it is reasonably correlated with the dependent or criterion variable. Another problem is that once the criterion variable is "adjusted" for pretreatment differences, it no longer exactly represents the behavior of the experimental subjects in the given situation. In other words, an assumption must be added that the adjustment will not jeopardize the interpretation of these results as representing some population of interest to the researcher. Finally, it is always better to randomize the assignment of subjects to treatment groups then to depend on some subsequent type of adjustments.

Other Variations

If we want to extend our basic analysis-of-variance design to include more than one dependent variable, this is possible through the use of the more complex *multivariate analysis of variance* (not included in this text; see Kerlinger and Pedhazur, 1973). If we wish to assess the degree to which a group of variables can aid us in differentiating among individuals or experimental treatments, *multiple discriminant analysis is used* (Chapter 15). Or as discussed in Chapters 9, 10, 11, and 12, relationships among independent and dependent variables can be assessed by use of *correlation* or *regression* methods.

Usually, however, the most frequent question about extending analysis of variance involves the use of more than one independent variable in a design, the topic of the next chapter.

SUMMARY

Single-factor analysis of variance is a statistical model used for testing the significance of difference among two or more means when these means reflect the consequences of different levels of a single independent variable. The statistical logic of analysis of variance is incorporated in the F ratio, a ratio of *between-groups* variance to *within-groups* variance. Given a calculated value of F, this value is interpreted in a sampling distribution for its probability under the terms of the null hypothesis. If this probability value is equal to or less than the criterion set for statistical significance, the null hypothesis is rejected in favor of the research hypothesis.

Since analysis of variance tests only the overall hypothesis of differences among means, it is usually necessary to conduct subsequent tests between individual means, often called *follow-up* tests. Some special considerations of analysis of variance models include whether or not the independent variable is a *fixed* or *random* factor, the number of *levels* of the factor that are incorporated into the design, and whether or not *repeated measures* are involved. *Analysis of covariance* is a method by which known pretreatment intergroup variations can be mathematically removed from measures of posttreatment intergroup variations prior to conducting an F test.

SUPPLEMENTARY READINGS

Basic readings on one-way analysis of variance may be found in:

Ferguson, George A., *Statistical Analysis in Psychology and Education,* 4th ed. McGraw-Hill, 1976. Chapter 15.

Kerlinger, Fred N., and E. J. Pedhazur, *Multiple Regression in Behavioral Research.* New York: Holt, Rinehart and Winston, 1973.

McCall, Robert B., *Fundamental Statistics for Psychology,* 2nd ed. New York: Harcourt Brace Jovanovich, 1975. Chapter 11.

Nie, Norman H., and others, *Statistical Package for the Social Sciences,* 2nd ed. New York: McGraw-Hill, 1975. Chapter 22.

Weinberg, George H., and John A. Schumaker, *Statistics: An Intuitive Approach,* 3rd ed. Monterey, Calif.: Brooks/Cole, 1974. Chapter 20.

MULTIPLE-FACTOR ANALYSIS OF VARIANCE

8

In Chapter 7 we saw how analysis of variance was applied to test the significance of difference among the means of three groups that varied in terms of a single factor. As an independent variable, this factor had three levels, each corresponding to the particular type of cartoon that was viewed by the children. We now turn to models that incorporate more than one factor.

MULTIPLE-FACTOR MODEL

Suppose that a study has more than one factor. What this means is that the design incorporates more than one independent variable, and each of these variables (or factors) can have two or more levels of its own. Usually we call such approaches *factorial designs* and describe them in terms of the number of factors and the number of levels each factor has. Thus, for example, a 2 × 2 design (usually read as "two-by-two") incorporates two factors, each having two levels, or a 4 × 2 × 2 design has three factors, the first having four levels, the second having two levels, and the third, two levels. In a given design, factors and their levels define different subgroups in the experiment. Multiple-factor analysis of variance provides methods for testing if different subgroups, or various combinations of subgroups, represent different populations in terms of what is being measured as the dependent variable.

A SAMPLE STUDY

For purposes of illustration we shall again return to the type of study used as an example in the last two chapters. As before, the groups are intentionally small and the data are simulated for ease of presentation.

Problem

It is generally agreed that kindergarten-aged boys have a greater tendency to engage in aggressive play than do girls of the same age. In considering the finding (sample studies in Chapters 6 and 7) that viewing unpunished aggression in a televised cartoon will affect a child's tendency to engage in aggressive play, the foregoing raises a question about the sex of the child: Will boys and girls be similarly affected by viewing unpunished aggression in a televised cartoon, as compared with a cartoon showing no aggression?

The above question suggests the need for a two-factor experiment. One factor will be the distinction in the type of cartoon viewed; its levels will correspond to whether the cartoon is one showing no aggression or one depicting unpunished aggression. The second factor will incorporate levels making the distinction between male and female children viewing the cartoons. Again the dependent variable will be ratings of the children's behavior during a play period after viewing one or the other of the cartoons.

The experiment, then, has two independent variables, the combinations of which are shown in Table 8.1. This is a 2 × 2 model, or a design incorporating "sex × cartoons." One of the useful features of a factorial design is that it allows us to test a number of different hypotheses in a single study. Note the potentially different population means shown in Table 8.1. For one thing, we can assess the effects of the two independent variables, one at a time. These are called *main effects*. In other words, quite apart from the different cartoon levels, do the male (m) and female (f) children represent the same population in terms of their mean ratings? Or, apart from the sex of the child, do the children in the groups who saw the aggressive (A) cartoon represent a different population than the groups who saw the nonaggressive (N) version? Suppose that the two research hypotheses (R.H.) and their corresponding null hypotheses (N.H.) are

$$
\begin{array}{ll}
(\text{R.H.} - \text{sex}) & \mu_m \neq \mu_f \\
(\text{N.H.}) & \mu_m = \mu_f \\
(\text{R.H.} - \text{cartoon}) & \mu_N \neq \mu_A \\
(\text{N.H.}) & \mu_N = \mu_A
\end{array}
$$

Table 8.1   **Diagram of the Two-Factor Sample Experiment**

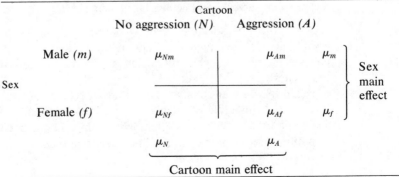

We say that these hypotheses are tested by assessing the *main effects* of the two independent variables.

Neither of the main effects hypotheses, however, directly addresses itself to the original question, that is, is there a differential effect that the two cartoons will have on boys as compared to girls? This requires that we somehow consider the effects of the two independent variables (cartoon, sex) in a combined form. Such a combined form in a factorial design is called an *interaction*. The nature of this interaction is the third hypothesis that can be tested in the present design.

Consider that our predicted interaction is that viewing an aggressive cartoon, as compared with a nonaggressive cartoon, will indeed have a different effect on boys as compared with girls. This prediction could be stated in the following mathematical form:

$$\text{(R.H.)} \qquad (\mu_{Nm} - \mu_{Am}) \neq (\mu_{Nf} - \mu_{Af})$$
$$\text{(N.H.)} \qquad (\mu_{Nm} - \mu_{Am}) = (\mu_{Nf} - \mu_{Af})$$

We say that the above hypothesis is tested by the interaction effects of the experiments.

Finally, suppose that the criterion set for rejection of each of the null hypotheses is $p < .05$.

Method

Using a kindergarten population as a pool of potential Ss for the experiment, two groups are initially selected according to sex. These groups constitute 10 males (m) and 10 females (f). Each of these groups of 10 is in turn randomly divided into two groups of five children each for the two different cartoon treatments (N, A). The four groups receive the following treatments:

Groups Am and Af see a 10-minute televised cartoon (A) in which the central character engages in aggressive behavior that goes unpunished.

Groups Nm and Nf see a 10-minute televised cartoon in which the central character shows no aggressive behavior (N).

As in the previous examples (Chapters 6 and 7), the dependent variable comprises judges' ratings of the children's tendencies to engage in aggressive behavior during a play period following the cartoon viewing.

When the data are analyzed, the means are as shown in Table 8.2. Notice that this table of means shows all of the means of the various subgroups that will be considered in testing the main effects and interaction effects of the study. The researcher uses a multiple-factor analysis of variance model as a method for deriving probability val-

Table 8.2   **Means for the Sample Experiment**

		Cartoon		
		No aggression (N)	Aggression (A)	
Sex	Male (m)	$M_{Nm} = 51$	$M_{Am} = 59$	$M_m = 55$
	Female (f)	$M_{Nf} = 45$	$M_{Af} = 49$	$M_f = 47$
		$M_N = 48$	$M_A = 54$	

Grand $M = 51$

ues for the three null hypotheses. For each of these null hypotheses an F ratio is calculated (Table 8.3), then each of these values of F is interpreted in an appropriate sampling distribution (Table 7.2) to determine its probability of occurrence under the terms of the null hypothesis. That is, it is asked in each case what the probability is of obtaining a particular value of F, if F is due to sampling error. For the present data, these F ratios and their associated probability values are as follows:

Cartoon main effect (testing N.H.: $\mu_N = \mu_A$): $F = 51.43$; with $d.f. = 1/16$, $p < .01$.
Sex main effect (testing N.H.: $\mu_m = \mu_f$); $F = 91.43$; with $d.f. = 1/16$, $p < .01$.
Interaction effects (testing N.H.: $(\mu_{Nm} - \mu_{Am}) = (\mu_{Nf} - \mu_{Af})$);$F = 5.71$; with $d.f. = 1/16$, $p < .05$.

Table 8.3   **Example of a Multiple-Factor Analysis of Variance (Skeletal Data)**

TOTAL SS (SUM OF SQUARES)
— where d_g equals deviation of scores about the grand mean (51).

Group Am	d_g	d_g^2	Group Nm	d_g	d_g^2
57	6	36	49	−2	4
58	7	49	52	1	1
58	7	49	52	1	1
61	10	100	53	2	4
61	10	100	49	−2	4
	$\Sigma =$	334		$\Sigma =$	14

Group Af			Group Nf		
47	−4	16	43	−8	64
50	−1	1	46	−5	25
50	−1	1	46	−5	25
51	0	0	47	−4	16
47	−4	16	43	−8	64
	$\Sigma =$	34		$\Sigma =$	194

Total SS $= \Sigma \, d_g^2 = 576$

Table 8.3 (*cont'd.*)

CARTOON MAIN EFFECT SS

— where d_c equals deviation of scores about the grand mean.

	Cartoons	
	A	*N*
Means	54	48
Deviation (d_c)	+3	−3
$d_c{}^2$	9	9
$nd_c{}^2$	90	90

$$SS = \Sigma\, nd_c{}^2 = 180$$

SEX MAIN EFFECT SS

— where d_s equals deviation of sex means from the grand mean.

	Sex	
	m	*f*
Means	55	47
Deviation (d_s)	+4	−4
$d_s{}^2$	16	16
$nd_s{}^2$	160	160

$$SS = \Sigma\, nd_s{}^2 = 320$$

ERROR (WITHIN) SS

— where d_w equals deviation of scores from group means.

Group *Am*	d_w	$d_w{}^2$	Group *Nm*	d_w	$d_w{}^2$
	−2	4		−2	4
	−1	1		+1	1
	−1	1		+1	1
	+2	4		+2	4
	+2	4		−2	4
Group *Af*			Group *Nf*		
	−2	4		−2	4
	+1	1		+1	1
	+1	1		+1	1
	+2	4		+2	4
	−2	4		−2	4

$$SS = \Sigma\, d_w{}^2 = 56$$

INTERACTION SS

where d_i equals deviations from the grand mean of means with main effects removed.

	Groups			
	Am	*Af*	*Nm*	*Nf*
Original means:	59	49	51	45
Cartoon effect removed:	56	46	54	48
Sex effect removed:	52	50	50	52
Deviation (d_i) from grand mean:	+1	−1	−1	+1
$d_i{}^e$:	1	1	1	1
$nd_i{}^2$:	5	5	5	5

$$SS = \Sigma\, nd_i{}^2 = 20$$

Table 8.3 (*cont'd.*)

SUMMARY TABLE

Source	SS	d.f.	MS	F*
Cartoons	180	1	180.0	51.43
Sex	320	1	320.0	91.43
Cartoon × sex	20	1	20.0	5.71
Error	56	16	3.5	
Total	576	19		

*MS of effect ÷ MS error.

Results

In the present statistical results there are grounds for rejecting all three null hypotheses, since in each case the calculated value of probability reached the level required for significance. Since the interaction hypothesis was of main concern in the study, the conclusion would center upon this outcome:

> There is evidence that the tendencies to engage in aggressive play after viewing a televised cartoon portraying aggression as opposed to a cartoon not portraying aggression are different for male and female children of kindergarten age.

Obviously, there is more to say about the results than is stated above. But such further interpretations will in most cases depend on subsequent comparisons of individual means (discussed later in this chapter).

F RATIOS IN A MULTIPLE-FACTOR MODEL

Most of the logic in calculating *F* ratios for the present multiple-factor model is not much different from what was discussed for the single-factor model. Again, the numerator of this ratio represents in each case a kind of between groups comparison, whereas the denominator is a variance estimate of the single population expected under the terms of the null hypothesis. Let us consider the nature of these sources of variation in the present example.

Total Variation

Just as in the calculation of *F* in Chapter 7, we can consider the concept of total variation of all scores about the grand mean. Examine this calculation in Table 8.3. The grand mean is 51. The total sum of squares, or the sum of all squared deviations of scores about the grand mean, is 576. This total sum of squares has "everything in the pot." Some of the variation represents main effects of the car-

toon factor; some of it comes from the sex factor. Still more variation may come from what special effects the two factors have when
combined, that is, interaction effect. Finally, some of the variation
represents individual differences among the subjects; this is referred
to as *error* variance. In the use of a multiple-factor analysis of variance, we shall be partitioning this total variation according to the
above sources. In the particular analysis model to be used, the variation expected under the null hypothesis, and the denominator of the
F ratios, is again the concept of variance *within* groups, or *error*
variance. Each of the other sources of variation, by contrast, will
contribute to the numerator of a series of F ratios—one for each
main effect, one for the interaction.

Main Effects

Note in Table 8.3 how the sources of variation due to the two main
effects (cartoons, sex) are calculated. Similar to the between-groups
source of variation in the single-factor model, each of these main effects reflects deviation of the two means of the levels of each factor
from the grand mean. Each main effect is treated as if the other main
effect did not exist. That is, for example, when the cartoon main effect is calculated, all ten scores for each of the two cartoon groups
(the two levels) are taken into account. There is no distinction made
for the fact that half of these scores came from boys and the other
half from girls. In short, the cartoon main effect is calculated as if
there were just two groups in the experiment, each one corresponding to a level of the cartoon factor. The comparison of cartoon
means ($M_A = 54$, $M_N = 48$) is then conducted by use of an F ratio
that has a main-effects variance in the numerator (180) and the error
variance (3.5) in the denominator. The main effect of sex is treated
in the same manner. The two levels of this factor ($M_m = 55$, $M_f =
47$) have a main-effects source of variation (320), and it too is cast
into an F ratio with the error variance (3.5).

Interaction Effects

Strictly speaking, variation due to interaction effects is only that
variation that is *not* attributable to either of the main effects nor
treated as error variance. That is, interaction variance is whatever
nonerror variation is observed among the individual groups (Nf, Nm,
Af, Am) when the main-effects variation has been removed. Put another way, interaction effects are not the sum of the main-effects
combination, but what nonerror effects occurred above and beyond
this sum. In other words, interaction centers upon the question:
What special additional effects are due to the combination of factors? This definition of interaction effects can best be seen by examining the calculation shown in Table 8.3.

Note the segment of Table 8.3 devoted to calculation of inter-action variance. Although this is not the typical way such variance is calculated (there are short-cut methods), it does illustrate just what is being considered in terms of interaction variance.

As we said above, the interaction variation is whatever remains in the way of group differences after the main-effects variations have been removed. Suppose, then, that we started with the means in terms of the actual outcome of the study; examine these in Tables 8.2 or 8.3. What are the consequences on the four individual group means if we remove the main effect of cartoon difference? Examine Table 8.4. In this case the two marginal means for cartoon effects have been equalized and, in turn, are equal to the grand mean. Making such an adjustment entailed reducing each of the individual group means for Cartoon A by 3 points, and increasing each of the group means for cartoon N by 3 points. The effect was to remove all difference among the means that was due to the cartoon factor. The consequences to the individual group means are shown in Table 8.4. Next we remove the main effects of the sex factor. This entails adjusting group means until the two sex means are equal to each other, as well as equal to the grand mean. The consequences of this upon individual group means are again shown in Table 8.4. What we have left are the group means (50, 52, 52, 50) which serve in the calculation of interaction variance.

If there were no interaction effects whatsoever (and no sampling error), we would expect each of these cell means to equal the grand

Table 8.4   **The Consequences of Removing Main-effects Sources of Variation**

Cartoon Effect Removed:

		Cartoon		
		N	A	
	m	54	56	(55)
Sex				
	f	48	46	(47)
		(51)	(51)	

cartoon means are now equal

Sex Effect Removed:

		Cartoon			
		N	A		
	m	50	52	(51)	
Sex					Sex means are now also equal
	f	52	50	(51)	
		(51)	(51)		

mean (that is, 51). But there are some differences remaining, and the question is whether their variation about the grand mean is significant or not. We calculate the variation (see Table 8.3) of these interaction means about the grand mean in the same way as the main-effects calculation. In the present example, the sum of squares is 20, as is the variance. This too is placed in an F ratio with the error variance and subsequently interpreted for a value of probability of occurrence under the null hypothesis.

Interpretations

The different components of variation are usually placed in a summary table similar to the one shown in Table 8.3. This table also reports the F ratios of interest and their associated probabilities. Again, the probability of each value of F occurring under the terms of the null hypothesis is gained by consulting a sampling distribution of F (for instance, Table 7.2).

Interpretation of main effects is relatively straightforward. In the present sample study, each main-effect comparison (for example, M_A versus M_N. M_m versus M_f) can be assessed from the outcome of the probability associated with the value of F. Since the probability was sufficiently low for rejection of the two null hypotheses, this in itself was enough to provide a basis for acceptance of the two main-effects research hypotheses and practical interpretation of the results. Had it been the case, however, that one or both factors had more than two levels, then it might have been necessary to make some subsequent statistical comparisons between individual means. Only by such comparisons would it be possible to find precisely what mean differences were contributing to an overall difference. These comparisons would entail much the same procedure as was described in Chapter 7 where there were three cartoons involved in the experiment.

The interpretation of an interaction does not necessarily stop with the appropriate F test. Suppose that, in the present case, the interaction had not been significant. Such an outcome would be clear-cut; we could not reject the null hypothesis posed for the interaction, hence would have no support for the hypothesized interaction. By contrast, however, since the interaction is significant, what kind of support does this give us? The significant interaction tells us that whatever differences are observed between cartoon levels A and N under one level of sex (for example, male) are not the same as found under the second level of sex (female). This in itself is a statement of the logical alternative to the null hypothesis; that is, the interaction research hypothesis that has been posed.

In order to see, however, if the outcome of the study conforms to our expected differences in terms of the relative effects of the car-

Table 8.5   Tabular Report of the Interaction Interpretation in the Sample Study

		Cartoon	
		N	A
Sex	m	51_b	59_c
	f	45_a	49_b

(Means with common subscripts are
not significantly ($p < .05$) different
from one another.)

toons upon boys and girls, it is necessary to examine further differences among individual group means. Such differences are tested by the use of the same methods of multiple mean comparisons as were discussed in Chapter 7. One test of this type[1] would lead to the outcome reported in Table 8.5 (calculations not shown). This puts the researcher in a position to state the results in considerably greater detail; for example:

> There is evidence that male children of kindergarten age have a greater tendency to engage in aggressive play than do female children of the same age when both have been exposed to a televised cartoon depicting aggression, as compared with a cartoon showing no aggression.

Interaction Patterns

Earlier we said that for the sample study a nonsignificant interaction would indicate that the difference in effects of the two cartoons (A, N) on males would be the same as the difference in effects of the two cartoons on females. By the same token, a nonsignificant interaction also says that the difference between the sexes (m, f) in effects of one cartoon (such as A) would be the same as the difference between the sexes in effects of the alternate cartoon (N). In reality, the foregoing two statements are simply two ways of saying the same thing in terms of the conditions that would underlie a nonsignificant interaction. To portray this point, consider Figure 8.1, two graphs of the same nonsignificant interaction. These two graphs illustrate a characteristic feature of a nonsignificant interaction: The lines are parallel, or near parallel.

By contrast, consider Figure 8.2, a graph of the significant interaction found in the sample study. Note how the lines are not parallel. The difference between cartoons A and N is greater for males (m) than for females (f). The fact, however, that an interaction is significant is not contingent on this particular pattern of differences

---

[1] D. B. Duncan, "Multiple Range and Multiple F Tests," *Biometrics*, XI (1955), 1–42.

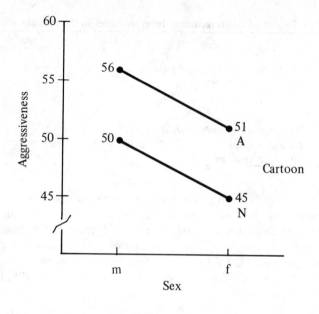

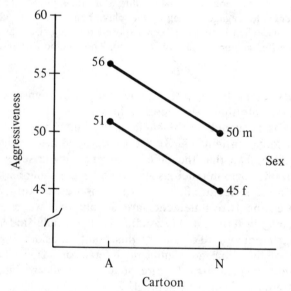

FIGURE 8.1   A nonsignificant interaction

among the individual group means. Consider, for example, Figures
8.3 and 8.4. Both of these illustrate at least two patterns that would
also underlie a significant interaction. In Figure 8.3, the pattern indicates an exact reversal of differences between the two cartoons for
each sex. On cartoon A the male group is lower in aggressive ten-

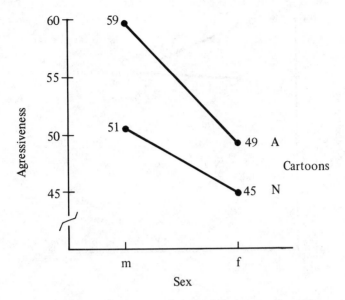

FIGURE 8.2   The interaction found in the sample study

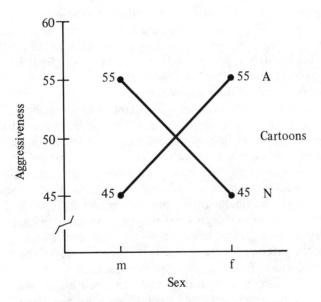

FIGURE 8.3   A significant interaction

dencies than on cartoon *N*, but just the opposite prevails for the female group. In Figure 8.4, the two cartoons have no differential effect on female *S*s, but the males show more aggressive tendencies after viewing cartoon *A* than after viewing cartoon *N*.

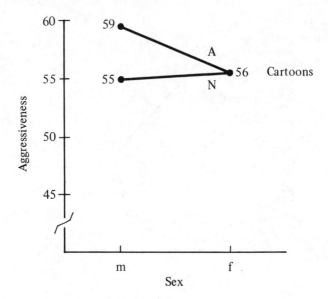

FIGURE 8.4   A significant interaction

   More patterns than shown here could illustrate cases of significant
interactions. The point in showing at least some of the different patterns is that a significant interaction, in itself, is usually not sufficient
as a basis for interpreting the detailed results. What is necessary is
an interpretation — often with subsequent comparisons of the individual means — of the interaction pattern.
   A final point concerns what consequences a significant interaction
has on interpretation of main effects. As we have been discussing all
along, a significant interaction indicates that whatever differences are
observed between (or among) levels of one factor are not consistent
across levels of another factor. Obviously, this often places conditions on how the main effects are interpreted. In the sample study,
for example, although male children consistently had a greater tendency to engage in aggressive play as compared with females, this
difference was greater after cartoon *A* than after cartoon *N*. Or consider the interaction portrayed in Figure 8.3. Here the main effects
of the cartoon factor as well as the sex factor would be nonsignificant (all means of the cartoon and sex main effects would be equal
to 50). Interpretation in this case of only the main effects would be
quite misleading. The experiment does indeed have significant effects, but these are solely in the interaction.
   The rule of thumb is usually this: If there is a significant interaction, then the interpretation of each of the main effects must take
into account the consequences of the interaction. On the other hand,
if the interaction is not significant, then the main effects can be
treated as independent of one another — that is, they can be inter-

preted individually, without being somehow conditional on one another.

FURTHER NOTES ON MULTIPLE-FACTOR MODELS

As in the discussion of single-factor models, we have done no more than touch the surface in describing a multiple-factor model. We can, however, introduce some additional points that are characteristic of more complex designs.

Levels and Factors

Like a single-factor model, the levels of individual factors in a multiple-factor design represent a compromise between theory and design. The sample experiment incorporated a 2 × 2 model—that is, two factors, each with two levels. Suppose, however, that we had incorporated four cartoons in the design, along with the two levels of sex. In this case we would have a 4 × 2 analysis model, that is, four levels of the cartoon factor and two levels of the sex factor.

Also, we can increase the number of factors in a design. Suppose that the sample experiment were expanded to incorporate not only the two levels of the cartoon factor and the two levels of the sex factor, but also two levels of a factor characterizing the time delay between viewing the cartoon and measures of aggressiveness in play. This would lead to a 2 × 2 × 2 factorial model. Moreover, suppose that two levels of a factor corresponding to the children's ages were also incorporated into the design. Here we would have a 2 × 2 × 2 × 2 model, that is, cartoon × sex × time × age. Finally, if we had four, rather than two, levels of the cartoon factor, and kept all of the above factors, the model would be described as a 4 × 2 × 2 × 2 design.

What we have seen as multiple-factor models has incorporated examples where for any given set of factors there are all possible combinations of the levels of each factor with the levels of another factor. Without going into detail, let us say that there are multiple-factor models that for one reason or another do not have this characteristic. One such design, called a *Latin square*, represents a systematic partial combination of levels of different factors. Latin squares are used mostly when a complete combination of all levels would be uneconomical. Another type of design (*nested*) represents a case where, for example, the two levels of a factor have nested within them different levels of another factor.

Higher-Order Interactions

Any time a factorial model incorporates more than two factors, it will incorporate an additional number of possible sources of interaction. Consider, for example, a 2 × 2 × 2 factorial model corresponding to factors of cartoon, sex, and time delay between viewing and measurement. The model for analysis would accommodate main-effects comparisons of the two levels of the cartoon factor, the two levels of the sex factor, and the two levels of the time factor. Interactions would include:

1. cartoon × sex
2. cartoon × time
3. sex × time
4. cartoon × sex × time

The first of these interactions (1) asks whether above and beyond the main effects of the cartoons and the sex factors, there is additional variation due solely to the combination of these two factors. Again, this is not the additive effects of cartoons and sex factors, but what unique effect their combination makes. Similarly, the second two of these interactions (2, 3) asks whether the combinations of cartoon and time factors, or sex and time factors, contribute to any unique effects. Finally, there is the three-way interaction (4), which asks if above and beyond any main-effects or two-way interaction effects, there is any special effect due solely to particular three-way combinations of the factors.

Needless to say, interactions become more and more complex, both in terms of computation and interpretation, as more and more factors are incorporated into a single design. In any two-factor design there will only be one interaction. But, for example, in a 2 × 2 × 2 × 2 model there are up to ten interactions that may be necessary to analyze.

Error Terms

Another key consideration in multiple-factor models is the basis for gaining the best estimate of error variance for a particular F ratio. Recall that this estimate of error becomes the denominator in a ratio with whatever source of variation is being subjected to an hypothesis test. In relatively complex models, the selection of the appropriate estimate of error (usually called the *error term*) becomes a relatively subjective decision. Within a given model for analysis, it is often the case that a number of different error terms will be employed, depending on what particular sources of variation are being tested. Such considerations as whether factors are fixed or random and whether measures are repeated or independent enter into the decision regarding the appropriate error term.

Other Variations

As is subsequently discussed in chapters on multiple regression (Chapter 12), and especially in those on multivariate statistics

(Chapters 13–15), there are common mathematical properties that relate analysis of variance to other statistical models, particularly multiple regression analysis. It is possible, for example, to take parts of the sample study in the present chapter and to calculate F ratios by another approach, multiple regression, one that centers on *relationships* between independent and dependent variables (Chapter 12). In fact, when you consider this alternative, you may see how many analyses of differences and analyses of relationships are close to reflecting the same thing: how one set of variables varies relative to another set.

Finally, as mentioned briefly in the previous chapter, analysis of variance can be done with multiple dependent variables, a technique known as multivariate analysis of variance (Kerlinger and Pedhazur, 1973).

SUMMARY

Multiple-factor analysis of variance is a statistical model for testing the consequences of manipulating two or more independent variables in a single research design. Each independent variable (factor) will have two or more levels. The F ratio is the statistic used to conduct the appropriate hypothesis tests in multiple factor designs. Significance tests among different levels of each factor are known as *main effects*. Whatever effects are due solely to the combination of factors are known as *interaction effects*. Given a significant interaction, it is usually necessary not only to conduct follow-up tests of means in order to interpret results, but it is also usually necessary to impose particular conditions on the interpretations of whatever main effects have been observed.

SUPPLEMENTARY READINGS

Readings on multiple factor analysis of variance and related topics may be found in:

Ferguson, George A., *Statistical Analysis in Psychology and Education,* 4th ed. McGraw-Hill, 1976. Chapters 16 and 17; 18, 19 and 20 on related topics.

Kerlinger, Fred N., *Foundations of Behavioral Research,* 2nd ed. New York: Holt, Rinehart and Winston, 1973. Part 5 (especially Chapters 14 and 15).

Kerlinger, Fred N., and E. J. Pedhazur, *Multiple Regression in Behavioral Research.* New York: Holt, Rinehart and Winston, 1973. Includes materials relating analysis of variance with regression analysis; also introduction to multivariate analysis of variance.

McCall, Robert B., *Fundamental Statistics for Psychology,* 2nd ed. New York: Harcourt Brace Jovanovich, 1975. Chapter 12.

Nie, Norman H., and others, *Statistical Package for the Social Sciences,* 2nd ed. New York: McGraw-Hill, 1975. Chapter 22.

Winer, B. J., *Statistical Principles in Experimental Design,* 2nd ed. New York: McGraw-Hill, 1971. Most of the volume is on analysis of variance designs.

NONPARAMETRIC TESTS

9

In addition to the tests discussed in the previous three chapters for assessing differences, there are tests that do not directly incorporate estimates pertaining to population characteristics. These are aptly called nonparametric *tests. In this chapter one such test, chi-square, will receive the most attention, although a number of other nonparametric tests of differences will be briefly mentioned.*

There is a variety of ways by which nonparametric tests can be classified. Perhaps it would be most practical to examine some of these tests in terms of the type of measurement typically involved in their usage. This latter point is one distinguishing feature of nonparametric tests. They are often used when measurement involves nominal or ordinal scaling.

WHEN DATA ARE IN CATEGORIES: USE OF CHI-SQUARE

As discussed in Chapter 2, there are times when measurement involves nothing more than assigning observations to different categories in a set of well-defined, mutually exclusive categories. Usually in this situation, called *nominal* scaling, we are either interested in comparing categories among themselves, or we are interested in contrasting how samples differ in terms of assignment into the categories. The typical statistical model used in such cases, and one having widespread utility in communications and education research, is chi-square. (Particular values of chi-square are usually identified by the symbol χ^2).

In essence, chi-square is best thought of as a discrepancy statistic. That is, its calculation is based on the discrepancy between the frequencies observed for a set of categories and some alternative theoretical set of frequencies posed by the researcher. Like t and F, χ^2 has a sampling distribution by which we can estimate the probability that a given value of χ^2 would be expected under the terms of the null hypothesis. Chi-square can be applied in a great variety of situations. We shall discuss some of the most typical of these.

Differences Among Categories

Sometimes chi-square is used not to compare different samples, but simply to investigate how items within a sample will distribute in a set of categories. This is usually called the *one-sample* case of chi-square.

Consider that a single sample of 30 teachers has been selected to evaluate three alternative versions (I, II, III) of a math lesson. Their task is to designate which one of the three lessons they consider to be the best planned. The measurement outcome of this study is designated in terms of the number of teachers judging each of the versions as best planned. These would comprise the *observed frequencies*. The researcher's hypothesis is that teachers will not choose equally among the three versions; that is, one or two versions will be selected over others as being the best. Such a research hypothesis implies the null hypothesis that teachers will choose equally among the three versions. The level for rejection of the null hypothesis is set at $p < .05$. Under the null hypothesis we would expect each version to have ten teachers rate it as the best planned; we call these the *theoretical frequencies*. Suppose that the following observations were obtained:

Version I: chosen by 4 teachers as best planned
Version II: 19
Version III: 7

As mentioned earlier, the basis for calculating χ^2 comes in the discrepancy between the observed and the theoretical distributions. The more is this discrepancy, the greater is the value of χ^2. Note in Table 9.1 how χ^2 is calculated for the present example. Here $\chi^2 = 12.6$, and when interpreted in a sampling distribution of chi-square (Table 9.2), this value is associated with $p < .01$. What this means is that the value of χ^2 would be expected in fewer than one of 100 cases of random sampling. Thus the researcher could reject the null hypothesis (no difference among the versions) in favor of the research hypothesis (a difference in choices of the versions).

Differences Among Samples

Consider a study similar to the one just described, but this time the research interest centers on whether males and females have differ-

Table 9.1   Calculation of χ^2; Single-Sample Case

	Versions			
	I	II	III	(total)
Observed freq.	4	19	7	30
Theoretical freq.	10	10	10	30

$$\chi^2 = \Sigma\left[\frac{(O - T)^2}{T}\right] = \frac{(4 - 10)^2}{10} + \frac{(19 - 10)^2}{10} + \frac{(7 - 10)^2}{10} = 12.6$$

With $d.f. = 2$, χ^2 of 12.6 is associated with $p < .01$ (see Table 9.2).

Table 9.2   **Example of a Table of χ^2 Sampling Distribution** *(See a complete table, p. 000)*

Probability	.10	.05	.02	.01
d.f.				
1	2.706	3.841	5.412	6.635
2	4.605	5.991	7.824	9.210
. . .	. . .	. . .	. . .	. . .
15	22.307	24.996	28.259	30.578
. . .	. . .	. . .	. . .	. . .
30	40.256	43.773	47.962	50.892

ent preferences in judging which of the three lessons is best planned. Rather than comparing the versions among themselves, males and females are now being compared in terms of their ratings of the three versions. Suppose that the research hypothesis holds that the two groups will have different judgments. The following data are obtained:

	Males	Females
Version I:	5	5
Version II:	5	20
Version III:	10	5

The above figures provide only the observed frequencies. Given the research hypothesis that the two samples (males, females) will be different, the implied null hypothesis is that they will not be different. The significance level of the chi-square test is set at $p < .05$. The theoretical frequencies, then, must represent this alternative, or null, hypothesis of no difference. The set of theoretical frequencies for the males and a (separate) set for the females must be calculated. This can be done by a relatively simple procedure (see Table 9.3) whereby each theoretical frequency is the product of the appropriate row sum times the comumn sum, divided by the grand sum. For example, the theoretical frequency for males choosing Version I is 20 (row sum) times 10 (column sum), divided by 50 (grand sum). These theoretical frequencies are entered in Table 9.3. For a given problem, each theoretical frequency accommodates differences among the categories but adjusts this to conform with the null hypothesis of no differences between the samples.

A value of χ^2 is then calculated on the basis of the discrepancies between each observed frequency and its corresponding theoretical frequency. In this case the total $\chi^2 = 9.03$ (Table 9.3). When interpreted in a sampling distribution (Table 9.2), this value is associated with $.01 < p < .02$. Thus the researcher rejects the null hypothesis (no difference between the samples) in favor of the research hypothesis (a difference between the samples).

Table 9.3   **Calculation of χ^2; Multiple-Sample Case**

		Versions		
	I	II	III	Row sums
Males:				
Observed freq.	5	5	10	20
Theoretical freq.	4	10	6	(20)
Females:				
Observed freq.	5	20	5	30
Theoretical freq.	6	15	9	(30)
Column sums	(10)	(25)	(15)	Grand sum = ((50))

$$\chi^2 = \Sigma\left[\frac{(O-T)^2}{T}\right] \frac{(5-4)^2}{4} + \frac{(5-10)^2}{10} + \frac{(10-6)^2}{6} +$$

$$\frac{(5-6)^2}{6} + \frac{(20-15)^2}{15} + \frac{(5-9)^2}{9} = 9.03$$

With $d.f. = 2$, χ^2 of 9.01 is associated with $.01 < p < .02$ (see Table 9.2).

Chi-square can apply to more than two samples in a single test. Thus, for example, a study might involve comparison of persons of five different experience levels in judging the aforementioned versions of a math lesson. Chi-square can also involve more categories than shown in the foregoing examples. Perhaps there are five or even ten versions of the math lesson; these could all be incorporated into a chi-square analysis. Or, for that matter, chi-square could apply to a design where five different experience level groups judged ten versions of the lessons.

In the sense just discussed, chi-square is often seen as a test of the goodness of fit between two distributions (considering distributions in terms of nominal measurement categories). Suppose, for example, that we have drawn samples of students from two universities and we are concerned with whether or not the two samples have students distributed equally in the same age categories. In other words, the question concerns the goodness of fit between the two distributions in terms of frequencies in various age groups. A nonsignificant chi-square would support the assumption that the two distributions fit one another, whereas a significant chi-square would indicate lack of fit.

Calculation of χ^2

The general formula for the calculation of χ^2 is given in both Tables 9.1 and 9.3. Just how the formula is applied is also seen in these two tables. For example, in the single sample use of chi-square, there were three cases of $O - T$. Each of these enters into the calculation of what appears between the brackets in the formula, then the results of the three cases are summed to yield a total value of χ^2. Similarly, in the example where two samples were compared in terms of selection from among the three versions of the article, all O - T cases are considered, only this time there are six cases to sum across in order to yield χ^2.

Interpretation of χ^2

Values of χ^2 are interpreted for values of associated probability in terms of a sampling distribution posed for the null hypothesis. In practice this sampling distribution is given in tabular form similar to the example in Table 9.2. Here it is necessary to enter the table with a value known for degrees of freedom ($d.f.$). In the case of chi-square for a single sample, $d.f.$ is equal to the number of categories considered minus one. Thus in the earlier example (Table 9.1), $d.f.$ is equal to $3 - 1$, or 2. When samples are being compared, $d.f.$ is equal to the number of samples minus one being compared times the number of categories minus one. In the foregoing example (Table 9.3), this was equal to $(2 - 1) \times (3 - 1)$, or 2.

As can be seen in Table 9.2, for given degrees of freedom, the larger the value of χ^2, the lower is the associated value of probability. This is a reflection of the fact that as the discrepancies between observed (O) and theoretical (T) frequencies increase, so does the value of χ^2, and the less is the probability of occurrence under an hypothesis of no difference.

WHEN DATA ARE RANK ORDERED: OTHER NONPARAMETRIC TESTS

In Chapter 2 the concept of *ordinal* scaling was introduced. In this scale we know nothing about the measurement intervals, but we do know that the categories have an ordered relationship. That is, we do know that one category represents a greater degree of some characteristic than another category. A very practical shortcoming when using an ordinal scale is that without a knowledge of measurement intervals there are limitations on the arithmetic operations that we can perform using the numbers assigned to these scales. This obviously imposes restrictions on the types of statistical models that can be used when ordinal scaling is involved, that is, unless the researcher is willing to hazard whatever error is involved in assuming that a scale has interval properties.

The statistical models that can be applied to ordinal scaling, in one way or another, stay within the limitations imposed by this level of measurement. In briefly reviewing some of these models, notice how they accommodate such limitations. Detailed descriptions of the calculations involved in the models to be discussed are found in Siegel's *Nonparametric Statistics*.

Independent Measures

We shall consider first the case where we have two groups of measures that are unrelated (that is, they are measures of different Ss). For purposes of comparison, we shall use the data from the sample study discussed in Chapter 6 where one group (1) of children was exposed to a televised cartoon depicting aggression, whereas a second group (II) saw a cartoon showing no aggression. The dependent variable comprises judgmental ratings of the children's tendency to engage in aggressive play after seeing one or the other of the cartoons. In the present case, however, we shall consider that the judges' ratings have only the power of an *ordinal* scale. It will now be necessary to assume that the average of judges' ratings for each child is now a *median* rather than a mean (since we cannot perform the arithmetic calculations necessary for a mean). These median ratings are as follows:

Group I	*Group II*
61	50
53	48
59	56
55	54
57	52

The hypothesis to be tested can no longer be stated in terms of mean scores; it will have to be adapted to the nature of the statistical test to be employed. However, the research hypothesis is still that the two groups will be differentially affected by the two cartoons. Some of the nonparametric tests that might be applied are next briefly discussed.

The *median test* can be used to test the research hypothesis that the two groups come from populations having different medians. The test itself, however, centers upon the null hypothesis that the groups come from populations having the same median. To understand how a probability value is calculated for this null hypothesis, examine how the data are displayed in Table 9.4. A median for all 10 of the scores (both groups combined) is found: in this case it is 54.5. The group scores are then distributed in a 2×2 table that indicates the number of scores in each group that are above this overall median, and the number of scores below this point. If the two groups had exactly the same median and there were no sampling error whatsoever, we would expect this division of scores to be the same for both groups. This would be the expectation under the terms of the null hypothesis. When a probability (p) value is calculated by use of an appropriate formula (Table 9.4), it indicates the probability that whatever discrepancies have been observed between the two groups could occur under the terms of the null hypothesis.

Table 9.4   **Example of the Median Test**

Grand Median of Combined distributions = 54.5

	Cartoons	
	I	II
	(A)	(B)
Ratings above grand median	4	1
	(C)	(D)
Ratings below grand median	1	4

Probability Calculation*

$$p = \frac{(A+B)!\,(C+D)!\,(A+C)!\,(B+D)!}{N!\,A!\,B!\,C!\,D!}$$

$$p = \frac{(5)!\,(5)!\,(5)!\,(5)!}{(10)!\,(4)!\,(1)!\,(1)!\,(4)!}$$

$$p = \frac{(120)\,(120)\,(120)\,(120)}{(3628800)\,(24)\,(1)\,(1)\,(24)}$$

$$p = .099$$

*"!" is the notation for a factorial. For example, 5! is read as "factorial 5" and means $1 \times 2 \times 3 \times 4 \times 5 = 120$.

As applied to the same data, the *Mann-Whitney U Test* does not test precisely the same hypothesis. The research hypothesis in this case says that the two groups come from populations having different distributions—or in terms of the null hypothesis, that the two groups have the same distribution. Table 9.5 illustrates how this test would be applied. The scores for both groups are ranked together from lowest to highest, while retaining a label to indicate what group each score came from. The value of U reflects the number of times scores for one group precede scores for the other group in this distribution. In the example in Table 9.5 no score for Group II is preceded by a Group I score until we reach the score 54; this contributes a value of one to the calculation of U. At the score 56, a Group II score is preceded now by two Group I scores. After the whole distribution is examined, the total value of U is seen to be 3. A probability value is then determined (usually by appropriate tables)

Table 9.5   **Example of the Mann-Whitney *U* Test**

Score Ranking:

Score:	48	50	52	53	54	55	56	57	59	61
Cartoon:	(II)	(II)	(II)	(I)	(II)	(I)	(II)	(I)	(I)	(I)
U:	0	0	0	0	1	0	2	0	0	0 = 3

U = the total number of times scores for Group I
precede scores for Group II

Probability of obtaining the above is <.056 (by use of tabled values).

that indicates the probability that a value of U of this magnitude or smaller could occur under the terms of the null hypothesis.

The *Kolmogorov-Smirnov Test* also tests the research hypothesis that two distributions differ. Table 9.6 illustrates how this test would be applied to the present data. Cumulative frequency distributions[1] of the scores in each group are prepared. The interval of greatest discrepancy between these two distributions then yields a particular value (two intervals in this case, both of them with a discrepancy of 3). A probability (p) value is then determined (by calculation or from an appropriate table) that indicates the probability of such a discrepancy occurring under terms of the null hypothesis.

Repeated Measures

When two sets of scores are somehow related (that is, measures on the same Ss, or matched Ss), different nonparametric models are employed. To illustrate two of these models briefly, consider that we have two sets of measures taken on the same six Ss. One set comprises measures (ordinal scale) of attitude toward blood donation taken prior to exposing the Ss to a 5-minute film encouraging such donation. The second set comprises postexposure measures. The research hypothesis, in a general form, is that the pre- and postexposure measures represent different conditions in terms of attitude toward blood donation (or more practically speaking, that the film has created a different condition of attitude). The ratings are as follows:

(S)	Preexposure	Postexposure
1	8	10
2	6	8
3	7	9
4	5	8
5	4	7
6	4	5

Table 9.7 illustrates how the *sign test* could be applied to the above data. Each rating pair is assigned either a plus (+) or a minus (−), indicating in this case whether the postexposure score exceeds the preexposure score (thus is marked +). The null hypothesis poses that the two sets of ratings are the same. If this were the case, we would expect roughly an equal number of pluses and minuses to occur. Or put another way, as the two conditions differ, we would expect to find a preponderance of one or the other sign. The value of

---

[1] In working up from the lowest score interval (Table 9.6), a cumulative frequency distribution shows the sum of the score frequencies up to and including each particular interval.

Table 9.6   **Example of the Kolmogorov-Smirnov Test**

Cumulative Frequency Distribution:

Score Categories	Cartoons I	II	
61	5		
60	4		
59	4		
58	3		
57	3		
56	2	5	→ discrepancy = 3
55	2	4	
54	1	4	→ discrepancy = 3
53	1	3	
52		3	
51		2	
50		2	
49		1	
48		1	

Probability of obtaining the above is > .05 (by use of tabled values).

Table 9.7   **Example of the Sign Test**

Directional Sign of Preexposure versus Postexposure Ratings:

S	*Ratings*		*Sign*
	Preexposure	*Postexposure*	
1	8	10	+
2	6	8	+
3	7	9	+
4	5	8	+
5	4	7	+
6	4	5	+

Probability of obtaining the above is < .05 (by use of tabled values).

probability (p) in this case refers to the probability that we would expect a particular discrepancy between plus and minus signs to occur under the terms of the null hypothesis.

As shown in Table 9.8, notice that the *Wilcoxon matched-pairs signed-ranks test* takes into account the magnitude of the difference between rankings of the scores in the two distributions. Similar to the sign test, these differences (now actual values of rank differences) are given plus and minus signs. If the null hypothesis of no difference between the two conditions were the case, we would expect the sum of the plus differences to equal roughly the sum of the minus differences. Again, a value of probability (p) indicates the probability under the terms of the null hypothesis of obtaining a par-

Table 9.8   Example of Wilcoxon Matched-pairs Signed-ranks Test

Magnitude and Sign of Score Differences:

S	Ratings		Differences	Rank of Differences	
	Preexposure	*Postexposure*		+	−
1	8	10	+2	3.0	
2	6	8	+2	3.0	
3	7	9	+2	3.0	
4	5	8	+3	5.5	
5	4	7	+3	5.5	
6	4	5	+1	1.0	

Σ of less frequent rank $(-) = 0$

Probability of obtaining the above is $< .05$ (by use of tabled values).

ticular discrepancy between the sums of the plus and minus differences.

FURTHER NOTES ON NONPARAMETRIC TESTS

Other Models

What we have covered thus far are only some of the more frequently used nonparametric tests. There are a number of additional tests of differences that may be more appropriate for certain research designs than the tests discussed in the chapter. For one thing, there are nonparametric tests that may apply when the researcher wishes to assess differences among more than two groups. Typically the *Friedman two-way analysis of variance* is used when repeated measures are involved, and the *Kruskal-Wallis one-way analysis of variance* is used for independent samples. Finally, there are nonparametric models used for tests of relationship. One of these, Spearman *rho,* is discussed in Chapter 10.

Power

The concept of statistical power was previously introduced in Chapter 5. Again, what it reflects is the probability of rejecting a null hypothesis that is, in fact, false. The concept of power is particularly apropos when considering nonparametric models because they are considered as generally less powerful than their parametric counterparts. What this means in practical terms is that, for a given set of data, the probability of rejecting a false null hypothesis is usually slightly better when using a test such as *t,* as compared with, say, the *median test,* the *Mann-Whitney U test,* or the *Kolmogorov-Smir-nov test.* Or put in terms of error, there is usually a slightly greater chance for committing Type II error (failure to reject a false null hypothesis) when using a nonparametric test, all other factors being equal.

Some illustration of the above point can be seen by comparing the obtained probability (p) values reported in the present chapter with the probability value obtained with the t test on the same data in Chapter 6. The difference was that the probability values obtained from the use of the three nonparametric models were all slightly greater than that obtained by use of the t test. Thus, for example, if the level set for rejection of the null hypothesis is $p < .05$, the various tests applied to the same data would lead to different conclusions in some cases.

Although the argument over the slightly lower power of nonparametric tests is a sound conceptual point, it is not always a practical one. First of all, much depends on the kinds of error that the researcher is willing to tolerate. Thus, for example, it may be more to a researcher's advantage to tolerate what are perhaps minor doubts about meeting necessary assumptions of a given test of a parametric type, rather than to hazard the error possible with a nonparametric test. On the other hand, to a certain degree the power of a given nonparametric test can be increased by increasing the size of the samples used in the study. Sometimes, too, the nature of the measurement and its distributions simply dictates the use of a nonparametric test, and there are no other alternatives. However, the decision to employ a nonparametric statistical test, and even the particular type of nonparametric test, is not always an easy one. This decision, like many others in reasoning with statistics, often depends on weighing a considerable number of relevant factors.

SUMMARY

Nonparametric statistical models are generally used when the researcher is faced with the use of nominal or ordinal scaling, or when particular assumptions cannot be made about the nature of the populations being studied. With nominal scaling a commonly used statistic is *chi-square.* It applies in tests centering upon how observations in a sample distribute into different categories, as well as how two samples may differ in terms of their distributions. Chi-square is also often used in the sense of a test of goodness of fit between distributions. When ordinal scaling is involved, one distinction among different nonparametric tests is whether the two sets of measures are related or not. There are additional nonparametric tests for comparing more than two groups, as well as tests of relationships among sets of measures. Generally speaking, nonparametric tests are considered less powerful than their parametric counterparts.

SUPPLEMENTARY READINGS

Kerlinger, Fred N., *Foundations of Behavioral Research,* 2nd ed. New York: Holt, Rinehart and Winston, 1973. Chapter 16 includes a discussion of nonparametric methods.

Siegel, Sidney, *Nonparametric Statistics for the Behavioral Sciences.* New York: McGraw-Hill, 1956. A still-excellent compilation of nonparametric methods and rationale for use.

Weinberg, George H., and John A. Schumaker, *Statistics: An Intuitive Approach,* 3rd ed. Monterey, Calif.: Brooks/Cole, 1974. Chapter 19 reviews selected nonparametric methods.

RELATIONSHIP ANALYSIS
PART 3

CORRELATION

10

In Chapters 6 through 9 we were primarily concerned with assessing differences between populations based on what we knew about differences in sample distributions. In all such cases the question of difference was centered in terms of measurement of some particular variable. In the present chapter, and in the two subsequent chapters, we turn to a different type of question to be answered statistically: To what degree do two or more variables show interrelationships in a given population? One way to assess such relationships is accomplished by statistical procedures of correlation.

THE NATURE OF CORRELATION

What is meant by correlation? Suppose that we have two sets of measurements taken on the same people. One measure consists of scores on an intelligence test; the other measure is an assessment of reading ability. Presumably, intelligence and reading ability are two traits that would tend to vary together in a given population. That is, if one person has a high score on the intelligence test, it is likely that he or she would also have a relatively high score on the reading test. Similarly, a person with a low score in intelligence would be expected also to have a low score in reading. In this case we would say that the two tests show *positive correlation;* that is, the measures vary together.

By contrast, suppose that we again had two sets of measurements taken on the same people. One measure is again a score on a test of reading ability, but the second measure is the time that it takes a person to read a passage of a given length. Presumably, a person with a relatively high score in reading ability would spend less time in reading a passage than a person with a low score in reading ability. In this case the two measures would be said to vary inversely; that is, as one measure is high, the other is low. This would be a case of *negative correlation.*

Correlation, then, characterizes the existence of a relationship between variables. Although there may be many reasons for a relationship, correlation says nothing about these reasons. It indicates only that two or more variables vary together either positively or negatively. This may be because one variable is the cause and the other the effect, or perhaps they both vary together as a result of a third variable being the cause. Correlation itself only indexes the degree of relationship; this index is called a *correlation coefficient*.

Coefficient of Correlation

A correlation coefficient indexes two properties of a relationship. The first of these is the *magnitude* of the relationship, that is, the degree to which the variables vary together. The second is the *direction* of the relationship, whether the variables vary together (positively) or whether they vary inversely (negatively). The coefficient itself has the following range of values:

$$+1.0 \text{ (perfect positive correlation)}$$
$$0.0 \text{ (no correlation)}$$
$$-1.0 \text{ (perfect negative correlation)}$$

Thus, for example, if a correlation of $+.50$ is found, this generally means that there is some degree of positive relationship, but it is not a perfect one.

Significance of Correlation

In addition to interpreting the magnitude or direction of a correlation coefficient, there are methods for testing the significance of a given value of correlation. Here the research hypothesis is typically that two variables are related; that is,

$$(\text{R.H.}) \text{ correlation} \neq 0$$

Accordingly, the null hypothesis says that no correlation exists, and that whatever value of correlation is found is due to sampling error.

$$(\text{N.H.}) \text{ correlation} = 0$$

A SAMPLE STUDY

Again, the present data are simulated. Ordinarily, a correlation study would not be undertaken with a sample this small.

Problem

Research literature suggests that people who read a lot also tend to be good readers, as compared with people who spend very little time reading. The researcher wishes to test this statement and at the same time gain some index of the magnitude of the relationship that might be involved. There is no reasoning here concerning cause and effect, only an interest in the degree to which measures of reading ability and time devoted to reading vary together in a given population.

Two measures are defined for the study. One, defined as variable X, is a score on a reading aptitude test. The second, defined as Y, is the average number of hours that a person has spent in reading books during each week of a year. Using the symbol r to represent a coefficient of correlation, the research hypothesis is stated as:

$$r_{XY} \neq 0$$

That is, it is predicted that the variables X and Y have a correlation that differs from zero, in either a negative or a positive direction. The null hypothesis is

$$r_{XY} = 0$$

The significance level is set at $p < .05$.

Method

After obtaining a sample of ten adults, the researcher has them estimate the average number of hours they spend reading books each week (variable Y), then administers to them a standard test of reading aptitude (variable X). The data are reported in Table 10.1.

Using a statistical method called Pearson product-moment correlation, a value of the correlation coefficient r is calculated (solution shown in Table 10.1). In this case, $r_{XY} = .937$. For the present sample size, the researcher estimates (Table 10.1) the probability that an r of .937 could occur due to sampling error, that is, under the terms of the null hypothesis. This probability value is $p < .01$. That is, we would expect a coefficient of this magnitude to occur in fewer than one chance out of 100 on the basis of random sampling.

Results

Since the probability of obtaining $r = .937$ is less than the level set for rejection ($p < .05$), the null hypothesis is rejected in favor of the research hypothesis. The researcher concludes that the two measures — reading aptitude and time spent in reading per week — are indeed related; for example:

The number of hours that a person spends in reading books each week is likely to have a very high positive correlation ($r = .937$ in the present study) with a measure of reading aptitude.

Table 10.1   Example of Pearson Product-Moment Correlation and a Test of Significance: Two-Variable Problem (Simulated Data)

Calculation of r

	Raw Data		Deviations from Means				
Ss	X (Reading Aptitude)	Y (Hours)	x	y	xy	x^2	y^2
1	20	5	0	0	0	0	0
2	5	1	−15	−4	60	225	16
3	5	2	−15	−3	45	225	9
4	40	7	20	2	40	400	4
5	30	8	10	3	30	100	9
6	35	9	15	4	60	225	16
7	5	3	−15	−2	30	225	4
8	5	2	−15	−3	45	225	9
9	15	5	−5	0	0	25	0
10	40	8	20	3	60	400	9
Σ	200	50			$\Sigma xy = 370$	$\Sigma x^2 = 2050$	$\Sigma y^2 = 76$
Mean	20	5					

$$r_{xy} = \frac{\Sigma xy}{\sqrt{\Sigma x^2 \cdot \Sigma y^2}} = \frac{370}{\sqrt{2050 \cdot 76}} = \frac{370}{\sqrt{155800}} = \frac{370}{394.7} = +.937$$

Significance of r (N.H.: $r = 0$)

$$t = \frac{r\sqrt{n-2}}{\sqrt{1-r^2}} = \frac{.937\sqrt{8}}{\sqrt{1-.878}} = \frac{2.65}{\sqrt{.122}} = 7.59$$

t is interpreted in Table 6.2, using $d.f. = n - 2$. Given this interpretation, the r of +.937 would be expected to occur in fewer than one of 100 cases if due to sampling error (that is, $p < .01$); thus, the null hypothesis of $r = 0$ is rejected.

THE LOGIC UNDERLYING PRODUCT-MOMENT CORRELATION

Like many of the statistical concepts covered in previous chapters, correlation reflects reasoning incorporating the concept of how scores vary within a distribution. Just how this reasoning works can be seen by considering what goes into the formula for product-mo-memt correlation.

The Formula for r

Consider first the formula that was employed for the calculations presented in Table 10.1. The x and y symbols represent deviations of the X and Y scores about the means of their two distributions. Recall (Chapter 3) that such deviations are a characteristic of the dispersion of scores in a distribution. The more dispersed are the scores about the mean, the larger these deviations will be. We shall see that a particular ratio among these deviations in the two distributions is what underlies a correlation coefficient.

Consider next what the denominator of the correlation formula represents. The product of the summed squared deviations of the two distributions (that is, $\Sigma x^2 \cdot \Sigma y^2$) is the total amount of the squared deviations that the two distributions could have maximally in common. We can represent this value as the area in a square, similar to the concept of sum of squares. We have done this in part (A) of Figure 10.1 in terms of the data from the same study. The length of one side of this square is the value corresponding to the square root of this area; this root value in turn corresponds to what we calculated for the denominator of the correlation formula.

Similarly, the numerator in the r formula incorporates deviations found in the two distributions. The numerator (that is, Σxy) represents one side of a square that would characterize the actual amount of squared deviations the two distributions have in common. This is shown graphically in part (B) of Figure 10.1.

What we have, then, in the r formula is a ratio between how much score deviation the two distributions actually have in common (that is, Σxy) and the maximum amount of score deviation that they could have in common ($\sqrt{\Sigma x^2 \cdot \Sigma y^2}$). Or in terms of the graphic portrayals in Figure 10.1, r is the ratio of the side of the smaller square to the side of the larger square, as shown in part (C).

We can also consider correlation from the standpoint of the amount of actual variance that the two distributions have in common. This is equal to r^2, or, in terms of part (C) in Figure 10.1, it is the ratio of the area in the smaller square to the area in the larger square.

The Significance of r

Under the terms of the null hypothesis, we can ask what the probability would be of obtaining an r of the magnitude of the one found. One method of estimating this probability is with the use of the t distribution. Using the procedure shown in Table 10.1, the value obtained for r is entered into a special formula for t. Given a value of t, it is interpreted for an associated probability level by consulting a sampling distribution of t (usually in tabular form such as Table 6.2). The degrees of freedom for entering the table are the number of pairs of scores, minus 2 ($N - 2$). Given the present example, with $t = 7.59$, and with $d.f. = 8$, and going back to Table 6.2, we see that p falls far below the .01 level. This tells us that with a null hypothesis of $r = 0$, if we repeated this correlation problem 100 times, we would expect to find an r of .93 less than one time if it were due only to sampling error.

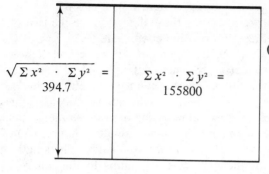

(A)   Square reflecting the denominator in the correlation formula used in the sample problem

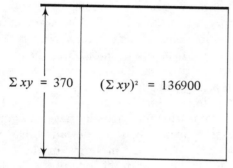

(B)   Square reflecting the numerator in the correlation formula used in the sample problem

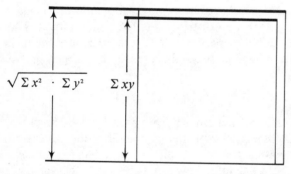

(C)   Ratio between the sides of the two squares which defines the correlation coefficient

FIGURE 10.1   Graphic portrayal of correlation coefficient

The fact that the correlation is statistically significant, however, does not mean that it is necessarily meaningful to us. What must be done next is to interpret its magnitude and direction.

The Interpretation of r

We have seen that there are three things to consider when given a particular value of r. First, there is the magnitude of the coefficient — it may vary from a zero or negligible level to a level that may approach perfect correlation (1.0). Second, there is the direction indicated by the sign of the coefficient. Do the measures vary directly (positively), or do they vary inversely (negatively)? Third, as discussed in the preceding section, there is the question of whether the obtained r is significant or not. That is, under the null hypothesis of $r = 0$, what is the probability of obtaining the value of r that was actually found? If this probability is sufficiently low (that is, below the level set for rejection), we would reject the null hypothesis in favor of the research hypothesis that the two measures are indeed correlated.

In most practical situations we would probably consider the foregoing three points in just their reverse order. That is, before considering the direction and magnitude of a coefficient, we would first of all want to know whether it is significant or not. If we are not in a position to reject the null hypothesis of $r = 0$, we would simply report this outcome as the sole interpretation of r. There would be no particular utility in talking about direction and magnitude of an r that perhaps represents only sampling error.

Given an r that is significant, its interpretation depends substantially on the nature of what is being studied. The direction of the coefficient, of course, is expected to conform with whatever expectations the researcher holds. If such expectation is fulfilled, the interpretation of direction is a straightforward matter.

By contrast, the interpretation of the magnitude of a correlation is often a subjective matter. Practically speaking, a relationship that might be considered moderate in one situation might be considered relatively negligible in another. Here we are talking about the psychological implication of the coefficient, not simply its magnitude in the statistical sense. For example, we might be interested in knowing what factors are related to a person's ability to write news stories. Knowing that, for instance, a person's grades in college English composition classes are correlated at the level of about $r = .42$ with ratings of news-writing ability seems of moderate consequence. The magnitude, although low in a statistical sense, is of sufficient psychological consequence that we might take this factor, among others, into account when considering what is related to news-writing ability. But on the other hand, suppose that our purpose were to have a good prediciton index of news-writing ability. The correlation of .42 is relatively negligible for this task. We would probably look for other factors as a basis for prediction. Thus, even though a correlation of .42 might be significant statistically, it may still be of relatively little consequence in terms of the problem under study. This last point perhaps needs special emphasis: Even though a correlation is statistically significant, its psychological significance remains to be interpreted by the researcher. In other words, statistical significance only puts researchers in a position to interpret the magnitude of the correlation; it does not do this job for them.

Even though the verbal description of a correlation remains highly contingent on what is being assessed, it is still useful to have some consistency of terminology in describing the magnitude of the coefficient itself. Although there is not a great deal of consistency in the research literature, Guilford[1] has suggested the following as a rough guide:

> $< .20$ slight; almost negligible relationship
> $.20 - .40$ low correlation; definite but small relationship
> $.40 - .70$ moderate correlation; substantial relationship
> $.70 - .90$ high correlation; marked relationship
> $> .90$ very high correlation; very dependable relationship

The Nature of r as an Index

One final point is a warning that a correlation coefficient says nothing about a percentage of relationship. That is, a correlation of .40 is not twice the relationship of one of .20. The coefficient itself is simply a convenient index, not an actual measurement scale.

In order to gain some feel for what an r implies, it is often useful to consider what r^2 signifies. As discussed earlier, r^2 is the proportion of variance that two measures have in common. In contrast to r, the value of r^2 can be thought of in terms of a proportion or percentage of relationship. When used for this purpose, r^2 is called a *coefficient of determination*. In terms of a coefficient of determination, we can say that two measures with an r^2 of .50 have twice the variance in common as two measures where $r^2 = .25$.

Figure 10.2 illustrates what we mean by having a feel for what r implies in terms of the statistical relation between two measures. Consider this graph of the relationship between r and r^2. Notice that relatively low values of r (for instance, .25) indicate an almost negligible amount (for example, $r^2 = .06$) of common variance between two measures. By contrast, note that as r becomes relatively large (for instance, .90), the corresponding proportion of common variance ($r^2 = .81$) is closer to the value of r; there is a substantial overlap in the variance of the two measures.

Another useful way to depict what r implies is shown by the overlap among circles as illustrated in Figure 10.3. Note that for the varying values of r, we can see r^2, or common variance, in terms of the overlap between the circles.

Not only is the coefficient of determination, or r^2, helpful in con-

[1] J. P. Guilford, *Fundamental Statistics in Psychology and Education* (New York: McGraw-Hill, 1956), p. 145.

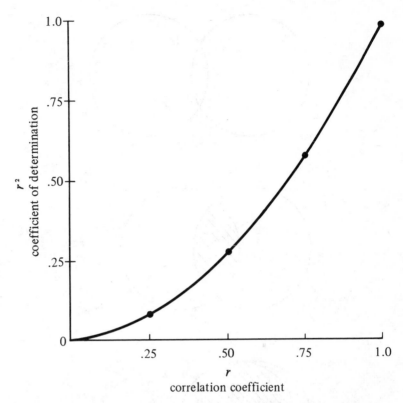

FIGURE 10.2   The relationship between correlation coefficients (r) and coefficients of determination (r^2)

sidering magnitude of correlation, it becomes particularly useful when we are considering correlation with reference to predicting values of one measurement from one or more other measurements, a topic treated in Chapter 11. Further, the idea of common variance among measures is a key concept in the logic of factor analysis, the topic of Chapter 13.

MULTIPLE CORRELATION

Thus far we have considered correlation where only two variables have been involved. Consider the case where we have three variables, X, Y, and Z, as shown in Figure 10.4. One way to describe the relations among X, Y, and Z is to consider them in terms of a series of two-variable problems. That is, we could calculate r_{XY}, r_{XZ}, and r_{YZ}, or r^2_{XY}, r^2_{XZ}, and r^2_{YZ}, and thus account for the relations shown in the three overlapping circles. We would probably report the results of the r calculations in the form of a table called an *intercorrelation matrix*, such as presented in Table 10.2. Here we can

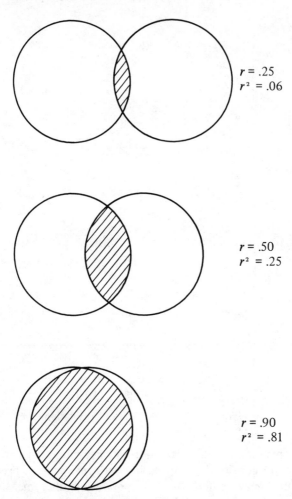

$r = .25$
$r^2 = .06$

$r = .50$
$r^2 = .25$

$r = .90$
$r^2 = .81$

FIGURE 10.3   Graphic portrayal of common variance (shaded area) associated with different values of r

specify the correlation of each of the three variables with each other. Obviously, such a matrix could be extended to accommodate the intercorrelations among any number of variables.

Rather than assessing the interrelations among more than two variables as a series of two variable problems, it is possible to assess the degree to which two or more variables together relate to a given variable. This is a problem in what is called *multiple correlation*.

Table 10.2   Example of a Matrix of Intercorrelations for a Three-Variable Problem

	Y	Z
X	.50	.40
Y	—	.45

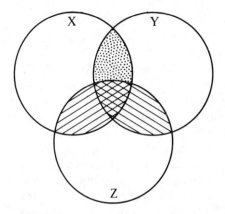

FIGURE 10.4   Graphic portrayal of a three-variable problem

Coefficient of Multiple Correlation

As an index of the relation of two or more variables with a given variable, the *coefficient of multiple correlation* is symbolized by R. The coefficient R has the same scale as the two-variable coefficient, r, that is,

+1.0 (perfect positive correlation)
 0.0 (no correlation)
−1.0 (perfect negative correlation)

The R symbol usually carries subscripts that indicate which variables are seen as related to a given variable. Suppose that Z is the variable focused upon, and R indexes the multiple correlation of X and Y with Z. The subscripts would be as follows:

$$R_{Z.XY}$$

whereas $R_{X.YZ}$ would symbolize the multiple correlation of Y and Z with X, or $R_{Y.XZ}$ would symbolize the multiple correlation of X and Z with Y. Similar to r^2, we can consider R^2 as a coefficient of multiple determination. The common variance defined by $R^2_{Z.XY}$ can be seen as the shaded area in Figure 10.4, that is, the area that variables X and Y have in common with variable Z. (You might consider also how $R^2_{X.YZ}$ and $R^2_{Y.XZ}$ would be shown in Fig. 10.4.)

A Sample Study

Consider a study where the researcher wants to assess the relationship between how individuals rate television programming (variable X) and the price paid for television sets (variable Y), with the average number of hours that they spend watching television during a week (variable Z). This defines a three-variable problem that will center on the calculation of $R_{Z.XY}$. Suppose that the number of individuals (N) surveyed equals 100.

The necessary data for calculating R are given in Table 10.3. These data comprise what was calculated (calculations not shown) as the individual correlation coefficients among the three variables — r_{XY}, r_{YZ}, and r_{XZ}. Using the formula given in Table 10.3, R_{ZXY} is calculated. In rounded form, it equals .97.

A coefficient of multiple correlation is interpreted in much the same manner as was r in a two-variable problem. Although we shall not go into the details of calculation, an obtained value of R can be tested against the null hypothesis that $R = 0$. This yields a value that can be interpreted to tell us the probability that the obtained multiple R could occur by sampling error. If this probability value is below that set for rejection, we are in a position to accept the research hypothesis that the variables X and Y have a significant multiple correlation with the variable Z. If the multiple correlation coefficient (R) is significant, then we would interpret its direction and magnitude much the same as was the case with r.

BINARY OR "DUMMY" VARIABLES

It is possible to estimate the correlation of continuous variables (e.g., interval scale) with categorical (nominal scale) ones by using *binary* (or *dummy*) variables where classes are coded by uses of "0" and "1." Suppose, for example, that in the earlier sample study (Table 10.1) in addition to the variables of reading aptitude (X) and

---

Table 10.3   **Example Calculation of a Coefficient of Multiple Correlation (Simulated Data)**

Given:

$$r_{XY} = .750$$
$$r_{YZ} = .918 \qquad N = 100$$
$$r_{XZ} = .894$$

Solution:

$$R^2_{Z.XY} = \frac{r^2_{XZ} + r^2_{YZ} - (2r_{XZ}r_{YZ}r_{XY})}{1 - r^2_{XY}}$$

$$= \frac{(.894)^2 + (.918)^2 - 2(.894)(.918)(.750)}{1 - (.750)^2}$$

$$= \frac{(.799) + (.843) - (1.231)}{1 - (.563)}$$

$$= \frac{.411}{.437} = .9405$$

$$R_{Z.XY} = \sqrt{.9405} = +.9698$$

hours spent reading books (Y), there was a further one of respondent sex (Z). Also, let us assume that the first five Ss were male, so their score of variable Z is each "0," and the second five are female, where $Z =$ "1" (or the numbers for the sexes could be reversed).

Using the formula for two-variable correlation, we would find that, as before (Table 10.1):[2]

$$r_{XY} = .937$$

The same formula would also indicate that:

$$r_{YZ} = .351$$
$$r_{XZ} - .387$$

Sometimes the dummy-variable approach is useful when we want to index the relationship of a group or treatment variable with a dependent or differentiating measure. Recall, for example, in Chapter 6 the example of two groups of children where one (I) had seen a cartoon showing aggressive activities, the other (II) saw a nonaggressive one, and the dependent variable was a measure of aggressiveness of the children's play activities after seeing one or the other cartoon. In the calculational example for t (Table 6.1), it was shown that the difference in means on the aggressiveness measure was statistically significant ($.02 < p < .05$). Thus it was concluded that seeing an aggressive cartoon as compared with a nonaggressive one was related to aggressiveness in the children's behavior. Given this outcome, we would expect to see a correlation between group membership (coded "0," "1") and the aggressiveness measure. Given Group I coded as "0" and Group II as "1," the correlation of this coded group variable with the aggressiveness measure is

$$r = -.66$$

If more than two categories must be coded, *each* can be included as a separate dummy variable. For example, to extend the above example, as in Chapter 7, there could be three groups: one seeing a cartoon where aggression occurs but punishment follows, another seeing unpunished aggression, and a third, a cartoon with no aggression. A child's membership in *each* group could be doded as "0" (not in) or "1" (in).[3]

---

[2] There are more precise methods for calculating correlations involving binary variables, but the present formula suffices for this illustration.
[3] Actually, only two of the three groups need be coded to do the analysis, because lack of membership in two implies membership in a third.

PARTIAL CORRELATION

It is also possible to describe the correlation of one variable with another while ruling out that degree of correlation attributable to other variables. Consider, for example, the three-variable graphic portrayal of correlation in Figure 10.4. The overall correlation of X with Z would be the total that the two circles overlapped. The correlation that both X and Z have in common with Y is the center cross-hatched area. If this were removed from the overlap between X and Z, the remaining section (diagonal lines) would represent the variance involved in the partial correlation ($r_{XZ.Y}$) of X with Z, with the joint effects of Y nullified.

For a numerical example, consider the intercorrelations summarized in Table 10.3 for the variables:

$$X = \text{ratings of television programming}$$
$$Y = \text{price paid for set}$$
$$Z = \text{time spent watching TV each week}$$
$$r_{XY} = .750 \qquad r_{YZ} = .918 \qquad r_{XZ} = .894$$

Suppose, then, that we wished to know the correlation of television ratings (X) with time spent watching TV (Z) while controlling for price of set (Y). Calculations are summarized in Table 10.4:

$$r_{XZ.Y} = .783$$

The result indicates a slight drop from the original correlation (.894), indicating that much of the XZ relationship is independent of joint relations with Y.

NONPARAMETRIC APPROACHES TO CORRELATION

The most common nonparametric alternative to Pearson's product-moment coefficient is a method known as *Spearman's rho*, some-

Table 10.4   **Example Calculation of a Coefficient of Partial Correlation**

Given:

$$r_{XY} = .750$$
$$r_{YZ} = .918$$
$$r_{XZ} = .894$$

Solution:

$$r_{XZ.Y} = \frac{r_{XZ} - r_{XY}\, r_{ZY}}{\sqrt{1 - r_{XY}^2}\ \sqrt{1 - r_{ZY}^2}}$$

$$r_{XZ.Y} = \frac{.894 - (.75)(.918)}{\sqrt{1 - .75^2}\ \sqrt{1 - .918^2}}$$

$$r_{XZ.Y} = .7834$$

Table 10.5   Example of Spearman Rho Correlation (Skeletal Data from Table 10.1)

	Raw Data		Ranking Values		D	
Ss	X (Reading Aptitude)	Y (Hours)	x_r	y_r	$(x_r - y_r)$	D^2
1	20	5	6	5.5	0.5	0.25
2	5	1	2.5	1	1.5	2.25
3	5	2	2.5	2.5	0.0	0.00
4	40	7	9.5	7	2.5	6.25
5	30	8	7	8.5	−1.5	2.25
6	35	9	8	10	−2.0	4.00
7	5	3	2.5	4	−1.5	2.25
8	5	2	2.5	2.5	0.0	0.00
9	15	5	5	5.5	−0.5	0.25
10	40	8	9.5	8.5	1.0	1.00
						Σ 18.50

Formula:

$$\rho = 1 - \frac{6\Sigma D^2}{n(n^2 - 1)}$$

$$= 1 - \frac{6(18.50)}{10(100 - 1)}$$

$$= 1 - .112$$

$$\rho = .888 \text{ (or .89)}$$

times called *rank-difference* correlation. This method is subject to less error than the product-moment formula when samples are relatively small (say, < 30) and also when measurement has only the power of an ordinal scale.

Table 10.5 illustrates how rho (ρ) would be calculated for the data presented in the sample two-variable study. Note first that the actual values entered into the formula are based on rankings of the two sets of values. For each pair of ranking values x_r, y_r, a difference (D) value is obtained. These difference values are squared, then summed and entered into the formula for ρ. Generally, ρ is interpreted in much the same way as a Pearson *r*.

Another type of rank-order correlation method is *Kendall's tau*. Although its calculation is different from rho, the tau coefficient has many similar characteristics and is interpreted accordingly.

SUMMARY

Correlation characterizes the relationship between variables—that is, the degree to which two variables vary together (*positive* correlation) or vary inversely (*negative* correlation). The *Pearson product-moment* correlation coefficient, *r,* has a range of values from $+1.0$ (perfect positive correlation) through -1.0 (perfect negative correlation). In addition to the direction and magnitude of r, we can also test the null hypothesis of $r = 0$ against a research hypothesis of $r \neq 0$.

The calculation of r reflects a ratio between the maximal amount of variability that two measures could have in common and the amount that they actually have in common. The coefficient squared (r^2) is the actual proportion of variance that two measures have in common.

A coefficient of multiple correlation (R) indexes the degree to which two or more variables are related with a given variable. The coefficient R is interpreted similar to r.

Correlation with a categorical variable can be calculated where dummy or binary ("0," "1") coding represents the categories.

Partial correlation provides a distinction of a relationship between two variables while controlling for effects of a third variable.

Selected nonparametric correlational models include *Spearman's rho* and *Kendall's tau*.

SUPPLEMENTARY READINGS

Basic readings on correlation may be found in:

Levin, Jack, *Elementary Statistics in Social Research,* 2nd ed. New York: Harper & Row, 1977. Chapter 11.

McCall, Robert B., *Fundamental Statistics for Psychology,* 2nd ed. New York: Harcourt Brace Jovanovich, 1975. Chapter 6.

Nie, Norman H., C. Handlai Hull, Jean G. Jenkins, Karin Steinbrenner, and Dale H. Bent, *Statistical Package for the Social Sciences,* 2nd ed. New York: McGraw-Hill, 1975. Computer analysis for correlation (Chapter 18) and partial correlation (Chapter 19).

Siegel, Sidney, *Nonparametric Statistics for the Behavioral Sciences.* New York: McGraw-Hill, 1956. Chapter 9 covers nonparametric approaches to correlation.

REGRESSION

11

In the previous chapter we were concerned mainly with the relation, or "co-relation," between or among variables. The correlation coefficient, for example, simply indexed the degree to which two or more measures varied together. Or the coefficient of determination was seen as indexing the amount of variance that two sets of measurements had in common. Obviously, knowledge of the relation among variables can be more than an end in itself. Sometimes we shall want to take advantage of a relationship and use it as a basis for prediction. That is, given a knowledge of the variable X and its relationship with the variable Y, how can we take particular values of X and predict what corresponding values of Y would be? The main statistical tool for answering this question is called regression analysis.

A SAMPLE STUDY

For purposes of illustration, let us consider the two-variable problems presented in the preceding chapter where variable Y was the average number of hours a person spent reading books during the week and variable X was a person's score on a reading aptitude test. When we use regression analysis, we are essentially interested in the description of a predictive relationship. Let us assume that we want to predict persons' times spent in reading books on the basis of knowing their reading aptitudes. We can calculate this prediction if we have an array of scores for the two variables or even if we simply have their means, the standard errors of those two means, and the correlation between the two variables.

Problem

Given that we know from the sample study in Chapter 10 (Table 10.1) that a correlation of .937 has been found between measures of reading aptitude and the average number of hours that people spend in reading books, we are interested in calculating a formula that will provide us with predicted values of the latter given the knowledge of values of the former. We are also interested in the reliability of our estimates.

Method

Taking the same scores from the sample of ten adults described in Table 10.1, and using a statistical method called *linear regression*, the following results are calculated (these calculations are subsequently described in Table 11.1):

$$Y' = .181X + 1.38$$

In this formula, Y' is the predicted number of hours that a person will spend reading books during each week of the year whereas X is whatever score we have for that person on the reading aptitude test. With this formula, when we take a given value of X, multiply it by .181 and add 1.38, we have our predicted value of Y. Suppose, for example, that for a given individual the reading score (X) is 40. As applied in this formula the results would represent a prediction (Y') of 8.62 hours per week. Or a reading aptitude score of 15 would predict 4.10 hours of reading per week, and so on.

How good is this estimate? Using an appropriate formula (discussed later) we can calculate what is known as *standard error of the estimate* (SEE). This will tell us that we can expect a particular amount of variation in our prediction given a particular probability. For example, we could conclude that the probability is 68 percent that our estimate will be within +.96 or −.96 of the value of Y that we calculated.[1] Or we could say that for a probability of 80 percent we could expect our estimate not to vary outside of +1.23 or −1.23. From these two examples, you can see that the larger the range of allowance in our estimate, the greater is the probability that the true prediction is *within* that range, or put in terms of error, the less is the probability that the true prediction is *outside* that range.

We can also test the statistical significance of the regression equation by use of an F test. In this case (calculation given later), $F = 57.33$, which we can interpret as statistically significant ($p < .01$.).

Results

A practical conclusion to this analysis, stated in mainly verbal terms, might be as follows:

The number of hours that a person spends in reading books each week can be quite accurately predicted from a measure of reading aptitude. A statistically significant prediction equation was

Reading time = (.181 × reading scores) + 1.38

With 20% probability of error, we can assume that our estimates will be accurate within +1.23 or −1.23 of the actual estimated figure.

[1] This SEE is relatively large, due in part to the small size of sample used for illustrative purposes.

THE LOGIC OF REGRESSION ANALYSIS

The term regression does not help much in introducing one to the logic of this type of prediction analysis. It is more use useful to consider how a line on a graph can describe the relationship between sets of scores on two variables. This is called a *regression* line, or *line of best fit*.

Line of Best Fit

Suppose that we have two variables, *P* and *Q*, and that we have four pairs of such scores.

	(P)	(Q)
Pair 1	10	7
Pair 2	20	12
Pair 3	30	17
Pair 4	40	22

We have plotted these scores in the graph shown in Figure 11.1; each pair is located in terms of the intersection of its *P* and *Q* scores

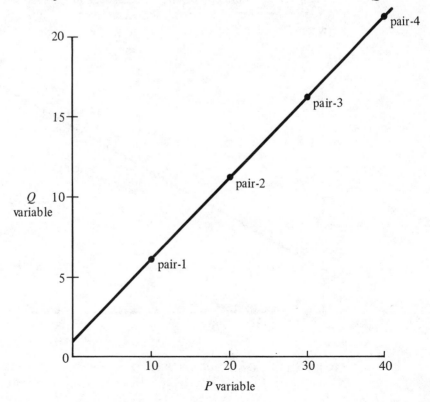

FIGURE 11.1   Distribution of pairs of *P* and *Q* scores and line of best fit

on the graph. Notice that if we connect these points with a straight line, we have a basis for predicting further corresponding pairs of P and Q scores. Thus, for example, if the relation between P and Q is indeed what we have observed, then if $P = 15$, $Q = 8.5$, if $P = 25$, $Q = 13.5$, or if $P = 38$, $Q = 20$, and so on.

The line in Figure 11.1 is a line of best fit or regression line. It defines our basis for predicting values of Q, given values of P (and vice versa in this particular example).

Suppose that we now plotted the pairs of scores from the sample study on a similar graph, as in Figure 11.2. Here we have the scores for each of the ten Ss plotted in terms of the intersection of each pair of X and Y values. Notice first that we cannot connect these points with the same kind of straight line as in the previous example. This is because the relation between X and Y values is not perfectly consistent as it was with the pairs of P and Q scores. Put another way, although we know that X and Y values tend to vary together (from their correlation calculated in Chapter 10), we do not have a perfect positive relationship ($r_{XY} = .94$, whereas $r_{PQ} = 1.0$).

Even though the relation between X and Y is not perfect, we can still employ the concept of the line of best fit. This time, however, the regression line is located such that given the observed relation between X and Y, we can attempt to make predictions of Y which,

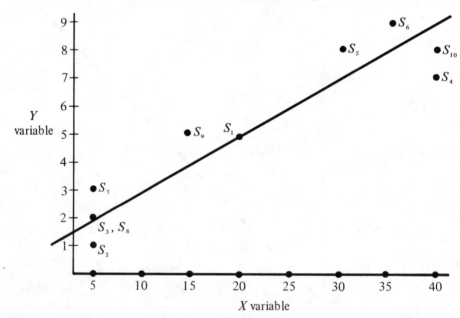

FIGURE 11.2 Distribution of X and Y scores from the sample study, including a line of best fit for the regression of Y on X (calculations in Table 11.1)

although not perfect, would involve the least degree of prediction error. In other words, our regression line is now a compromise in getting the line of best fit. This line then serves as our basis for predicting Y values, given various values of X.

Our computational procedure for obtaining this line will be discussed at a later point. The present line applies only for predicting Y values, given X values, and not the reverse. (The reverse would involve location of a slightly different line, one giving the best estimate of X values, given various values of Y.) The present approach also assumes that a straight line will be the best basis for prediction, a procedure known as *linear regression*. There are alternative procedures for problems involving nonlinear regression (see Supplementary Readings at the end of the present chapter).

Regression Equation

We usually describe the line of best fit in terms of a mathematical formula called a regression equation. This equation tells us the best-fitting mathematical linear relation between X and Y values. That is, given a value of X, it tells us what must be done to derive the best estimate of a value of Y. This equation is actually the mathematical expression for a straight line, an equation that you may have met in elementary algebra. Its general form is

$$Y = bX + a$$

For a given line, a defines what is called its *intercept,* whereas b defines its *slope*. Notice in Figure 11.1 that as the line approaches the smaller values of P and Q, it intersects the Q axis at 2.0. This says that in terms of intercept (and we must still consider slope), the variable Q has values 2.0 units greater than corresponding values of P. Or, in Figure 11.2, note that the intersection on the Y axis is at about 1.38, indicating that in terms of intercept, corresponding points along Y exceed points along X by 1.38. By contrast, slope (or b) tells how many units the variable Y is increasing for every unit increase in X. For example, in Figure 11.1, b is equal to 0.5, which means that (in addition to intercept) for every unit increase of P there is a 0.5 unit increase in Q. Similarly in the case of Figure 11.2, b is equal to .181, indicating that for every unit increase in X, Y increases .181 units. Imagine what would happen to the slope of the line if b were increased. It would become steeper when an increase in a unit of X would have a greater increase in the corresponding unit of Y. Or if b were decreased, increases in a unit of X would be associated with lesser increases in a unit of Y.

Generally speaking, a and b are usually referred to as *regression coefficients* (or sometimes *constants*). Given the knowledge of what *a* and *b* are equal to in a given problem, we can then employ the above equation to estimate a value of *Y*, given a particular value of *X*. (We indicate that *Y* is an estimated value by giving it a prime sign, that is, *Y'*, or when estimating *Q*, we write *Q'*.)

For some practical examples in the use of this equation, consider first the perfect relation between *P* and *Q* shown in Figure 11.1. Given that $a = 2.0$ and $b = 0.5$, then:

$$\text{If } P = 10, \text{ then } Q' = (.5 \cdot 10) + 2 = 7, \text{ or}$$
$$\text{If } P = 20, \text{ then } Q' = (.5 \cdot 20) + 2 = 12, \text{ and so on}$$

Similarly, for the line of best fit shown in Figure 11.2, if $a = 1.38$ and $b = .181$ then:

$$\text{If } X = 10, \text{ then } Y' = (.181 \cdot 10) + 1.38 = 3.19, \text{ or}$$
$$\text{If } X = 40, \text{ then } Y' = (.181 \cdot 40) + 1.38 = 8.62, \text{ and so on}$$

In order to see again the relation between the above equation and the line of best fit, note how the estimated values of *Y* (*Y'*) and values of *Q'* fall on their respective lines of best fit (Figs. 11.1, 11.2) at the intersection of appropriate *X* and *Y* values (or *P* and *Q* values).

Calculation of *a* and *b*

The essence of the calculation in the foregoing formula is to determine what *a* and *b* are equal to for a given problem. Table 11.1 provides an appropriate formula and illustrates how this formula would be employed for calculations applicable to the sample study where *X* is a person's score on a reading aptitude test, and *Y* is the average number of hours a person would spend in reading books during a week. Notice that the equation appears in a composite form; that is, *a* and *b* are both calculated with the single formula shown in Table 11.1. Generally, what we have described as *b* corresponds to the $r \left(\dfrac{S_Y}{S_X} \right)$ portion of the formula, and $(X - M_X) + M_Y$ corresponds to *a*. As shown in Table 11.1, the solution gives us the following equation:

$$Y' = .181X + 1.38$$

which is the same as discussed earlier, and which describes the line of best fit shown in Figure 11.2. The foregoing calculations were our basis for determining this line.

In practical terms, again, we could say that:

$$\text{Reading time} = (.181 \times \text{reading scores}) + 1.38$$

Table 11.1   Example Calculation of the Regression Equation for the Prediction of Y Values from X Values (Simulated Data from Table 10.1)

Composite formula:

$$Y' = r_{XY} \left(\frac{s_Y}{s_X}\right)(X - M_X) + M_Y$$

where Y' is the predicted value of the Y variable
    r_{XY} is the correlation between X and Y ($r_{XY} = +.937$)
    s_X and s_Y are the standard deviations of the X
        and Y distributions ($s_X = 14.31$; $s_Y = 2.76$)
    X is a given value of the X variable
    M_X and M_Y are the sample means of the distributions
        of the X and Y variables ($M_X = 20$; $M_Y = 5$)

Calculations:

$$Y' = .937\left(\frac{2.76}{14.31}\right)(X - 20) + 5$$
$$= (.181)(X - 20) + 5$$
$$= (.181X) - 3.62 + 5$$
$$Y' = .181X + 1.38$$

We would use the above equation in order to make whatever predictions that we desired, or we might construct a graph such as shown in Figure 11.2.

Standard Error of the Estimate

As described earlier in the sample study, the *standard error of the estimate* (SEE) provides us with an estimate of the accuracy of our prediction (or error, depending on how we look at it). Note in Figure 11.2 how the actual values of Y scores (S_1, S_2, . . ., S_{10}) vary about the regression line. As you would logically expect, the greater that these values vary from the ones potentially based on the regression equation (i.e., the ones on the regression line), the greater is the probability of error in making predictions from that equation (or line). Error increases with the decrease in linear correlation between the two variables. Note the typical equation below for calculation of SEE for the prediction on Y from X:

$$SEE_{YX} = s_X \sqrt{1 - r^2_{YX}}$$

If you examine the components of this equation, you will see that the greater is the correlation of X and Y (r_{YX}), the less will be the standard error of the estimate (SEE). You may also note that the greater is the variability or standard deviation of the X distribution (s_X), the greater will be the standard error of the estimate.

As implied by its label, SEE is interpretable as a standard error score on a normal curve. More practically, the calculation tells us what a error score (of *one*) is equal to in terms of the Y scale of scores. It is thus interpretable in terms of probability values. Recall in Chapter 4 how *one* standard error of the mean encompassed 34 percent of a normal distribution on each side of a mean, or 68 percent, counting both. In the case of SEE, we see what one standard score is equal to in our predicted score values; then, by interpreting a normal curve table, we can make such statements that if SEE = .963 we would expect the true prediction of Y to fall inside plus or minus 1 SEE of the estimate of Y 68 percent of the time, or 32 percent outside of it.

A simple way to visualize standard error of the estimate is to plot it along with a regression line, and this has been done in Figure 11.3 for the sample study. The regression line is the same as in Figure 11.2 but now we have added a dashed line above and below the regression line to represent the 34 percent probability that estimates of Y may be higher than predicted by the regression line, and the 34 percent probability that they will be less. Figure 11.3 is a useful reminder that in using regression equations we are making estimates, all with calculable margins of error.

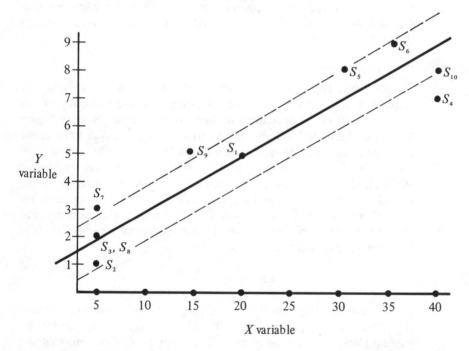

FIGURE 11.3   Range of deviation of Y scores one SEE above and one SEE below the line of best fit for sample study and where SEE = .963

Application of the F Test

There are several strategies for testing the statistical significance of a simple regression analysis. One way is to test the significance of the correlation coefficient such as described earlier in Chapter 10, pp. 124–126.

Another, and more often used, method is by use of an F test, an approach not unlike what was described in the Chapters (7 and 8) on analysis of variance (see especially pp. 80–83.) In fact, one reason it is useful to understand how F is used in regression is that it reveals how regression and analysis of variance are based on the same underlying concept of dividing total variance into components. There is variance that is "accounted for" (such as *between groups* in analysis of variance) and variance that is "unaccounted for" (such as *within groups*) in the sense that it is attributable to sampling variability. Recall (pp. 80–81) that an F ratio was

$$F = \frac{\text{Variance between groups} \div \text{degrees of freedom}}{\text{Variance within groups} \quad \div \text{degrees of freedom}}$$

Given a calculated value of F and the degrees of freedom associated with the components of the F ratio, we can then make probability estimates that the F occurred by chance. We can test hypotheses with this F.

In regression analysis, F essentially represents

$$F = \frac{\text{Variance predicted (or "regression")} \div \text{degrees of freedom}}{\text{Variance unpredicted (or "residual")} \div \text{degrees of freedom}}$$

Logically, the more that the predicted variance exceeds the unpredicted variance, the less we would expect the value of F to have arisen by chance.

The components for F, like in analysis of variance, are drawn from sums of squares. The numerator, or predicted variance, is the sum of squared deviations of the predicted variable about its mean, or simply the sum of squares due to the "regression." The denominator is the sum of squared deviations of differences between the known values of the predicted variable and the predicted values, or simply the sum of squares of the "residual." (Total variance would be the sum of squared deviations of the known values of the predicted variable.)

For the sample study of the prediction of hours spent in reading books (Y) from scores on a reading aptitude test (X), $F = 57.9$. The degrees of freedom for this F are the number of variables minus one, and the size of sample, minus the number of variables minus one, minus one. F is interpreted in a table of F (as in Table B in the Appendix). Here we find the F of 57.9 to be statistically significant ($p < .01$). We could thus reject a null hypothesis that our predictions were no better than chance.

F tests are particularly useful in regression equations with more than one predictor variable ("multiple regression," Chapter 12). We can, for example, test whether each additional predictor variable contributes significantly when added to the equation.

FURTHER NOTES ON REGRESSION

Use of Binary or Dummy Variables

In Chapter 10 (pp. 132–133) we illustrated how categorical variables could be entered in a correlational analysis by using "0" and "1" scores to code group or treatment membership. Recall how data from the sample study in Chapter 6 (Table 6.1) were used to calculate the correlation of a measure of a child's aggressive play with whether the child had just seen a cartoon depicting aggressive behavior (Group I, coded "0") or one showing no aggression (Group II, coded "1"). The correlation was $r = -.66$, indicating a relationship between cartoon exposure and aggression in play.

Similarly, binary or dummy variables can be entered as predictors (only) in a regression equation. Using the data from Table 6.1 and coding the groups as described above as the Y variable, and X as scores on aggressive behavior, the resulting equation is

$$X' = -5Y + 57$$

Here we would not be especially interested in using the formula as a basis for predicting values of the aggressive play scores (X) on the basis of treatment groups membership (Y). (Note how respective binary values of Y, 0 or 1, result in predicting the means of the two groups, 57 or 52.) Instead we could take our basis for calculating this equation and calculate an F ratio, which in this case is 6.25. At 1 and 8 degrees of freedom this is statistically significant ($p < .05$). In brief, F can be calculated in this way to test the statistical significance of a hypothesis of group (or treatment) differences.

Dummy variables can be used as above in the role of a substitute for a t test or analysis of variance of group differences. However, you will find them more often in communication and education studies where the researcher wishes to test the role of a nominal variable along with other variables in a problem of multiple prediction (Chapter 12).

Relationship with t Tests and Analysis of Variance

Several times we have noted that regression analysis and analysis of variance are generally two ways of looking at the same thing. They involve dividing total variance into "accounted for" components and estimating the probability that the variance accounted for is greater than could occur by chance alone.

If we ran an analysis of variance on the data in Table 6.1, it, like the binary regression analysis, would result in an $F = 6.25$. Similarly, to recall the relation between t and F ($F = t^2$) mentioned briefly in Chapter 7, the t of 2.50 reported in Table 6.1 also equals an F of 6.25. In short, t, F, and linear regression analysis are all very similar in their underlying components.

SUMMARY

In addition to indexing the relationship between variables, we can take this relationship as a basis for *prediction*. That is, given values of one variable, what are the best estimates of related values of another variable? The basis for such prediction is found in the logic of regression analysis. The predictive relationship can be shown in terms of a *line of best fit*, or in terms of the mathematical definition of this line, called a *regression equation*. We can calculate a range of probable error for our predictions, that is, the *standard error of the estimate* (SEE). Also, we can test whether or not predicted variance exceeds a chance occurrence by use of the F test (the same essentially as in analysis of variance).

SUPPLEMENTARY READINGS

Guilford, J. P., and B. Fruchter, *Fundamental Statistics in Psychology and Education,* 5th ed. New York: McGraw-Hill, 1973. Chapter 15 is a very clear introduction to the logic and computation of regression equations.

Kerlinger, Fred N., *Foundations of Behavioral Research,* 2nd ed. New York: Holt, Rinehart and Winston, 1973. Chapter 35 is a practical introduction to research designs involving linear regression.

Nie, Norman H., C. Handlai Hull, Jean G. Jenkins, Karin Steinbrenner, and Dale H. Bent, *Statistical Packages for the Social Sciences.* New York: McGraw-Hill, 1975. Chapter 20 describes the use of a packaged computer program for calculating linear regression equations, SEE, and F tests.

MULTIPLE REGRESSION

12

Just as we can assess the multiple correlation of two or more variables with another variable, it is also possible to bring more than one predictor variable to bear in predicting scores on a given variable. This technique is known as multiple regression analysis. It can be applied for practical purposes of predicting a score on a variable given scores on two or more predictor variables. It is also used as a method of describing the relative degree of contribution of a series of variables in the multiple prediction of a variable.

A SAMPLE STUDY

Problem

Recall in Chapter 10 the example where it was determined that the multiple correlation of a person's evaluation of television programming (variable X) and the price paid for the television set (variable Y) was +.97 (rounded) with the number of hours a person would spend watching television per week (variable Z). In the present sample study of multiple regression, we are interested in achieving four objectives: (1) developing a formula for practical prediction of variable Z from variables X and Y; (2) seeing if the predictions are better than chance; (3) estimating the accuracy of the predictions; and (4) describing the relative contributions of the variables X and Y to the prediction of Z.

Method

Using a calculation procedure (described subsequently in Table 12.1), we determine that the prediction formula is as follows:

$$Z' = (-10.13) + 2.51X + .076Y$$

Here Z' is the variable that we predicted, or the number of hours that a person would spend watching television per week. The variable X is whatever scaled rating we have of a person's evaluation of a television program, and variable Y is the price the person paid for the television set. Further calculations are made for F and standard error of the estimate.

Results

Given the foregoing formula as a basis, the results can be summarized as follows:

1. Viewing time $= (-10.13) + (2.51 \times$ programming rating) $+ (.076 \times$ price of set).
2. This prediction equation is statistically significant far beyond a chance level ($F = 750.0$, $d.f. = 2/97, p < .001$).
3. With a 32 percent margin of error, we know that the predicted viewing time is likely to fall within the range of $+2.9$ or -2.9 hours.
4. From further calculations we can determine that approximately 42 percent of the prediction of viewing time is accountable to programming ratings and approximately 52 percent of this prediction is due to price of set.

Thus, for example, we could say that if the average programming rating (variable X) was 4.0 for an individual and the price of set (variable Y) was \$180, then the estimated viewing time per week is 13.59 hours. Within 32 percent chance of error, we would estimate that this could vary between 16.49 and 10.69 hours. Or, for another example, we could say that with average programming ratings of 6.0 and a price of set equal to \$330, the predicted viewing time is 27.73 hours, with a 32 percent margin of error of our estimate being between 30.63 and 24.83 hours.

THE LOGIC OF REGRESSION ANALYSIS

Equation for Multiple Regression

Similar to the regression equation presented in Chapter 11, the equation for multiple regression incorporates regression coefficients. Again, there is the coefficient a; but this time we have two b coefficients, one for the X predictor variable, and one for the Y variable. The equation has the general form:

$$Z' = a + b_{XZ \cdot Y} X + b_{YZ \cdot X} Y$$

In this case the b coefficients are different for the two predictor variables. The coefficient $b_{XZ \cdot Y}$ is the ratio of how many units Z increases for every unit increase in X, but with the effects of Y held constant. By contrast, $B_{YZ \cdot X}$ is the ratio of how many units Z increases for every unit increase in Y, but where the effects of X are held constant. Because each of these two coefficients only reflects a portion of the prediction of Z, they are called *partial regression coefficients*.

The essence of the calculation is to find out what the a and the two b coefficients are equal to. As shown in Table 12.1, it is necessary to calculate first what are called *beta* (β) *coefficients,* or *standard partial regression coefficients,* prior to calculating the two b coefficients. These calculations are based on correlations found among the variables. The formulas for calculating coefficients are given in Table 12.1. Given the two β coefficients, the two b coefficients are then calculated. These calculations are also shown in Table 12.1. Calculation of the a coefficient incorporates the means of the three variables, as well as the two b coefficients. This calculation is also shown in Table 12.1 for the sample data.

Interpretation of the Multiple Regression Equation

Given the solutions for the a and the two b coefficients for the sample data, we have the following prediction equation:

$$Z' = (-10.13) + 2.51X + .076Y$$
$$(\text{or})$$

Viewing time $= (-10.13) + (2.51 \times \text{programming}$
$\text{rating}) + (.076 \times \text{price of set})$

The foregoing should hold, for example, for what we know about the means of the three variables; that is if $X = 5$ and $Y = 205$, then Z should equal 18. That is,

$$Z' = (-10.13) + (2.51)(5) + (.076)(205)$$
$$= (-10.13) + (12.55) + (15.58)$$
$$Z' = 18$$

Overall F Tests in Multiple Regression

The overall equation in multiple regression, as in the sample problem, can be tested for statistical significance. This is done by use of an F test, the same as discussed in Chapter 11 (pp. 000–000) for a single-variable prediction equation. It is also essentially the same as using an F ratio to test the statistical significance of the multiple correlation coefficient associated with the multiple regression equation.

Note in Table 12.2 how the formulas for the calculation of F for either the regression equation or the multiple correlation coefficient are what we described in Chapter 11 as the ratio of "accounted for" (or "regression") variance to "unaccounted for" (or "residual") variance. They result in essentially the same F (within rounding errors), which in the sample problem is statistically significant ($p < .001$). The practical interpretation is that we have a prediction that with very little probability could have arisen from chance alone. Or, put another way, the test is against the null hypothesis that $R = 0$.

Table 12.1   **Example Solution to Equation of the Multiple Regression of Z on X and Y (Simulated Data from Table 10.3)**

Given:

$r_{XY} = .750$	$M_X = 5$	$s_X = 2.2$
$r_{YZ} = .918$	$M_Y = 205$	$s_Y = 88.0$
$r_{XZ} = .894$	$M_Z = 18$	$s_Z = 11.8$
$N = 100$ cases		

Beta (β) coefficients:

$$\beta_{XZ \cdot Y} = \frac{r_{XZ} - r_{YZ}r_{XY}}{1 - r_{XY}^2}$$

$$= \frac{(.894) - (.918)(.750)}{1 - (.750)^2}$$

$$= \frac{.205}{.438} = .468$$

$$\beta_{YZ \cdot X} = \frac{r_{YZ} - r_{XZ}r_{XY}}{1 - r_{XY}^2}$$

$$= \frac{(.918) - (.894)(.750)}{1 - (.750)^2}$$

$$= \frac{.248}{.438} = .566$$

b coefficients:

$$b_{XZ \cdot Y} = \left(\frac{s_Z}{s_X}\right)\beta_{XZ \cdot Y}$$

$$= \left(\frac{11.8}{2.2}\right)(.468)$$

$$= 2.510$$

$$b_{YZ \cdot X} = \left(\frac{s_Z}{s_Y}\right)\beta_{YZ \cdot X}$$

$$= \left(\frac{11.8}{88.0}\right)(.566)$$

$$= 0.076$$

a coefficient:

$$a = M_Z - b_{XZ \cdot Y}M_X - b_{YZ \cdot X}M_Y$$
$$= (18) - (2.510)(5) - (.076)(205)$$
$$= (18) - (12.55) - (15.58)$$
$$= -10.13$$

Multiple regression equation:

$$Z' = a + b_{XZ \cdot Y}X + b_{YZ \cdot X}Y$$
$$Z' = (-10.13) + 2.51X + .076Y$$

Table 12.2   **Example Solution of an _F_ Ratio for the Multiple Regression Equation or Multiple Correlation Coefficient in the Sample Study**

For the equation

$$F = \frac{(\text{Sum of squares, regression}) \div (\text{number of independent variables minus one})}{(\text{Sum of squares, residual}) \div (\text{sample size minus number of independent variables minus one})}$$

$$F = \frac{12947.3397 \div 2}{837.42022 \div 97} = \frac{6473.6695}{8.633198} = 749.9$$

For the multiple correlation coefficient:

$$F = \frac{(\text{multiple correlation coefficient squared}) \div (\text{number of independent variables minus one})}{(\text{one minus multiple correlation coefficient squared}) \div (\text{sample size minus number of independent variables minus one})}$$

$$F = \frac{(.9691)^2 \div 2}{(1 - .9691^2) \div 97} = \frac{.4696}{.0006268} = 749.2$$

Standard Error of the Estimate

Just as in the simple linear regression example, the standard error of the estimate (SEE) reflects the standard deviation of actual values of the predicted variable (here, Z') from actual ones (Z). Calculation of SEE allows us to make a probability statement of the likelihood that an estimate will fall into some range of values about the predicted one. The standard error of the estimate reflects the following relationship:

$$SEE = \sqrt{\frac{\text{Sum of squares residual}}{N - k - 1}}$$

where

N = sample size
k = number of independent variables

Or, in the sample study:

$$SEE = \sqrt{\frac{837.42022}{97}} = 2.93 \text{ or } 2.9 \text{ (rounded)}$$

Contributions of Predictor Variables

We can also use the multiple regression equation to tell us how much each of the predictor variables contributes to the joint prediction of Z. Earlier (Chapter 10) it was noted that a squared multiple correlation coefficient tells us what proportion of the variance of the predicted variable is accounted for by the two predictor variables. In order to see the relative contribution of each predictor variable to this variance, we multiply each correlation coefficient of a predictor variable with the predicted variable (that is, r_{XZ}, r_{YZ}) by its respective beta (β) coefficient—for instance,

$$R^2_{Z.XY} = (\beta_{XZ.Y}r_{XZ}) + (\beta_{YZ.X}r_{YZ})$$

Employing the sample data gives us:

$$(.9691)^2 = (.468)\,(.894) + (.566)\,(.918)$$
$$(.94) = (.42) + (.52)$$

Thus, we can conclude that programming ratings contribute .42 to the prediction of viewing time, and price of the television set contributes .52 to this joint prediction.

FURTHER NOTES ON PREDICTOR VARIABLES

In communication and education research, the achievement of a high degree of prediction (that is, a large R^2) is not always an end in itself. We could be concerned with the selection or elimination of predictor variables that might be useful or not useful as "causal" descriptions. We could be interested in how the relationships between predictors and predicted variables may vary by the inclusion of other variables. Or we might have the purpose of testing one alternative set of predictors against another. In brief, it is helpful to be aware of some of the procedures by which predictor variables are evaluated. We have already seen one example in the last scetion. This was in how the total predicted variance (R^2) could be broken down into components of variance (product of predictor variable correlations times β coefficients) contributed by predictor variables. Following are some further considerations of predictor variables.

Evaluation of *b* Coefficients

It may be recalled from Chapter 11 that a *b* coefficient, in essence, defines the amount that a predicted variable should change for increments of a predictor variable (remember the *slope* in Fig. 11.1?). In multiple regression, *b* coefficients are called *partial* because they reflect only the relation of the predictor with the predicted variable with other variables held constant. These coefficients are thus one way of evaluating a predictor variable.

Although we shall not go into detail on the calculations, it is possible to calculate the standard error of a *b* coefficient. This is done in situations where the researcher is interested in the reliability of a predictor variable in a multiple regression equation. The coefficient will vary among samples, and calculation of its standard error provides an estimate of the amount of such variation.

By the same token, it is possible to test the significance of a b coefficient against the null hypothesis that in the population it is zero. This provides a statistical basis for deciding if a variable, relative to other predictors, has a significant contribution to the equation.

Other F Tests

There are several ways to test the significance of the amount of regression variance accounted for by a predictor or even a subset of predictors in a multiple regression equation. They all have the same logic of comparing "accounted for" ("regression") variance with "unaccounted for" ("residual") variance. Such tests might include the significance of adding or excluding one or several variables from an equation, the significance of one or a set of predictors compared with others, or changes in a predictor or subset of predictors given changes in other variables in the equation. All such tests are easy to conceptualize, and today many computer programs for multiple regression supply the basis for calculating them.

"Stepwise" Analyses

Using a number of modern computer programs for calculating multiple regression equations, it is possible to develop the equation one variable at a time until some criterion is reached which indicates that further predictors are unnecessary. This is called a "stepwise" procedure. Usually, the variables are selected in the order of their ability to contribute to the overall prediction. The criterion on whether or not to add an additional variable is often whether or not its contribution to predicted variance is statistically significant (say, by F test). An important practical note in stepwise analyses is that if additional variables are added to the equation this may change the relative contribution of variables added earlier. Some computational programs include the elimination of variables added earlier if a variable added later makes their contribution insignificant.

Emphasis on Predictor Variables Rather Than on R^2

One of the basic interpretations of regression equations is the degree of prediction of the dependent variable as characterized by a large R^2 and a low SEE. Another basic focus, and one coming into increasing use in communications and education research, is on an interpretable and practically useful configuration of predictor variables, perhaps even at the expense of maximizing R^2. Suppose, for example, that prediction of children's success in use of a "reading instruction" kit is based on IQ, sex, reading readiness score, teacher rating, time spent with the kit, and parental attitudes. Some of these predictor variables are more easily obtained (for example, sex) than others (parental attitudes), hence any multiple regression equation that emphasizes the former predictors, even if it yields a somewhat smaller R^2, will be more useful than one involving "costly" variables. Also, the researcher or administrator may wish to emphasize predictor variables that are themselves "manipulatible" in the classroom, such as time spent with the kit. As another example, we could probably increase the predictability of a student's success in graduate studies if we added additional predictors based on in-depth interviews. However, such interviews are not always feasible and are generally costly, so we settle for *Graduate Record Examination* scores (or another test), undergraduate grade point average, and letters of recommendation.

USE OF BINARY OR DUMMY VARIABLES

In both Chapter 10 (pp. 132–133) and Chapter 11 (p. 146), we described how categorical variables could be coded with "0" or "1" and entered into correlational or regression analyses, a procedure using "binary" or "dummy" variables. This procedure can be extended as well to multiple regression analysis. We shall discuss two applications of such variables of practical utility to the reader: (1) coding categorical questionnaire or survey data, and (2) coding independent variables so as to illustrate the similarity of multiple regression analysis to analysis of variance.

Categorical Variable Coding and Analyses

One of the statistical limitations of many questionnaire-survey studies is that data often include categorical or nominal scales (Chapter 2) that do not lend themselves directly to more powerful or complex statistical analyses. Examples of categorical variables often found in surveys are sex, ethnicity, living area, occupation, and type (rather than amount) of training. In many studies involving such variables, analyses are limited to simple frequency tables of "cross-tabulations." Suppose, for example, that we were interested in studying the relationship between categorical variables and level of newspaper readership as gauged in average number of hours spent reading a newspaper each day. Results could be presented in a variety of cross-tabulations such as:

		Sex	
		Male	Female
Occupation	Labor	1.2 hours	1.1 hours
	Management	2.1 hours	1.8 hours

Although such tabulations may be useful for simple descriptive purposes, they become complicated and cumbersome as more variables are added, especially continuous variables (such as age, income, and amount of education).

If the dependent variable is continuous (as is "hours" in the example), the relation of it with other variables, categorical or continuous, can be assessed in a multiple regression equation. This involves coding categorical variables, as described in Chapters 10 and 11, usually with "1" if a category applies, "0" if not. For example:

Variables			
Male	Female	Labor	Management
0	1	1	0

The above represents coding for a female laborer. Given such coding as predictor variables, it would then be possible to (1) to assess the overall predictability of newspaper reading from sex and occupation data, (2) to test the statistical significance of the relationship, (3) to calculate the SEE of the prediction, and (4) to estimate the relative contribution to the prediction of the two predictor variables. Moreover, it is not difficult to envisage adding further predictor variables, categorical or continuous.

Binary of dummy variables greatly enhance our ability to reason from categorical data.

VARIATIONS OF MULTIPLE REGRESSION ANALYSIS

Just as we have discussed the relationship between regression analysis and analysis of variance in Chapter 11, there are similar relations with several other statistical approaches, most of which come under the heading of *multivariate* statistics. For example, if categorical variables serve as intended dependent variables, a related statistical method is *multiple discriminant analysis* (Chapter 15). If the correlations between *groups* of variables are assessed, this involves a related method called *canonical correlation* (Chapter 14). Finally, there is the extension of multiple regression analysis to accommodate multiple dependent variables, a technique called *multivariate multiple regression* analysis (see Kerlinger and Pedhazur, 1973, or Monge and Cappella, 1978, in the Supplementary Readings at the end of this chapter).

SUMMARY

Multiple regression analysis provides for the prediction of values of a variable, given values of two or more variables. The equation describes the arithmetic combinations of the predictors for the best estimate of the predicted variable. The equation can be tested for statistical significance by use of an *F* test. The standard error (*standard error of the estimate* or SEE) of the predicted values can be calculated. The equation for multiple linear regression is an extension of the two-variable simple regression equation. Much of the focus in multiple regression analysis is centered on predictor variables—their statistical significance in regression equations, effects of adding or deleting variables, and their practical interpretation. Binary or dummy coding allow categorical variables to be introduced as predictors in multiple regression equations.

SUPPLEMENTARY READINGS

Guilford, J. P. and B. Fruchter, *Fundamental Statistics in Psychology and Education,* 5th ed. New York: McGraw-Hill, 1973. A classic in the field. Chapter 16 is clear and of practical value on multiple regression analysis.

Kerlinger, Fred N., *Foundations of Behavioral Research,* 2nd ed. New York: Holt, Rinehart and Winston, 1973. Chapter 35 contains a good overview of multiple regression.

Kerlinger, Fred N., and E. J. Pedhazur, *Multiple Regression in Behavioral Research.* New York: Holt, Rinehart and Winston, 1973. A "must" for any person working seriously with multiple regression and multivariate statistics.

Monge, Peter, and J. Cappella (Eds.), *Multivariate Techniques in Communication Research.* New York: Academic Press, 1978.

Nie, Norman H., C. Handlai Hull, Jean G. Jenkins, Karin Steinbrenner, and Dale H. Bent, *Statistical Packages for the Social Sciences.* New York: McGraw-Hill, 1975. Chapter 20 provides practical notes on multiple regression analysis as well as a guide for a packaged computational program.

MULTIVARIATE ANALYSIS

PART 4

FACTOR ANALYSIS

13

In Chapters 10–12 we examined various ways to characterize the degree and direction of interrelationship among two or more variables in a population. In discussing correlation and regression we were typically concerned either with relationships among individual variables or with the relationship of one variable to a group or cluster of variables. In the last 20 years or so behavioral scientists have made increasing use of statistical methods that provide for assessments of patterns within, among, or between clusters of variables. Such analytic methods are usually referred to as "multivariate analyses," particularly when a given analysis has more than one dependent variable.[1]

In this final section we shall introduce three multivariate methods:

1. *Factor analysis, which asks the degree to which clusters of intercorrelated variables may represent fewer underlying, more basic, hypothetical variables*
2. *Canonical correlation, an estimate of the correlation between two clusters of variables*
3. *Multiple discriminant analysis, which assesses the degree to which clusters of variables will significantly distinguish individuals or cases into two or more groups*

Because these methods are computationally complex and for all practical purposes require the use of an electronic computer, there will be less emphasis on their calculation as compared with methods in the earlier chapters. We shall, however, attempt to provide highly simplified illustrations of some of the main concepts underlying their calculation.

A PRACTICAL VIEW OF FACTOR ANALYSIS

In the three previous chapters we have seen how relationships among measures are assessed in terms of the variance that they have in common. We saw in Chapter 10 how common variance was the underlying concept in describing correlation—an index of the degree

---

[1] Some writers include multiple regression analysis as a multivariate method because of its close mathematical relationship to the methods discussed in Chapters 13–15.

and direction to which measures vary together. In Chapters 11 and 12 we saw how a knowledge of common variance between or among measures could be used in formulas for the prediction of measurement values of one variable, given a value or values of other related variables. In both correlation and regression we were concerned with what we might generally call *measurement overlap.*

In addition to assessing the correlation between measures, or predicting one measure from another, we can view measurement overlap as the extent to which measures of different variables are measuring something in common. Factor analysis is a mathematical procedure applied for this purpose. It aids us in answering the question: Given a relatively large number of variables, does their measurement overlap, indicating that there may be fewer, more basic, and unique variables underlying this larger number? Factor-analytic procedure takes the variance defined by the intercorrelations among a set of measures and attempts to allocate it in terms of fewer underlying hypothetical variables. These hypothetical variables are called *factors.*

A SAMPLE STUDY

Problem

Suppose, for example, that we have obtained a number of different measurements all presumed relevant in one way or another to the use of an educational television series and classroom activities both aimed at reducing sex-role stereotyping on the part of elementary-school students. Our aim is to determine if there is an interpretable underlying structure which can be found in the relationships among these variables.

Method

For purposes of illustration, let us simulate a study with the following variables:

1. Student's IQ as obtained from a standardized test.
2. *Male stereotype score,* a measure of the degree of male stereotyping a student imposes when estimating whether males, females, or both would fill a given occupation
3. *Female stereotype score,* the degree of female stereotyping in estimating who would fill an occupation
4. *Recall of TV program content,* scores on a multiple-choice test about the TV programs
5. *Comprehension of instructional objectives,* measuring the stu-

dents' understanding of the problems of imposing stereotypes as shown in the TV programs and activities

6. *Reasoning from instructional objectives,* the ability to draw inferences about the negative effects of stereotypes
7. *Students' liking of materials,* scaled ratings of the TV programs and activities
8. *TV viewing frequency,* number of programs viewed
9. *Classroom activity frequency,* number of exercises in which a student has participated
10. *Teachers' attitudes toward the materials,* scaled ratings by the childs' teacher of favorability of attitude toward the TV program and activities; this rating is assigned to each childs' set of measures

We would expect that a number of the foregoing measures would be interrelated ("intercorrelated"). For example, the different measures involving the understanding and reasoning from the program content would probably be interrelated, perhaps reflecting some overall comprehension aspect. Similarly, we might expect that variables such as TV viewing frequency or classroom activity frequency would be related to some of the outcomes of the study, such as the attitude or comprehension measures. Recall how in Chapter 10 we graphically portrayed interrelationships in terms of overlapping circles showing common variance (Figs. 10.3 and 10.4). Using this same system of circles, we might picture the overlap among the foregoing ten variables as shown in Figure 13.1.

The overlap among circles (Fig. 13.1) can be thought of as the proportions of variance that the various measurements have in common, or the square root of each of these proportions of common variance could be depicted as correlation coefficients as summarized in Table 13.1. We can look at the circles in Figure 13.1 or the correlation coefficients in Table 13.1 measure by measure if we wish, and see what each measure has in common with each other measure. From a broader viewpoint, however, we might consider how all ten measures tend to cluster about one another. We can think of each cluster, or any circle not in a cluster, as perhaps reflecting some aspect of the behavior that we are assessing. That is, our measures of recall, comprehension, and reasoning do indeed seem to be measuring roughly the same thing—something that we might more generally call basic comprehension. We can note also how the two different measures of stereotype bias in occupations, one for males and one for females, might simply be representing an overall bias toward stereotyping. By the same token, we can see in Figure 13.1 or in Table 13.1 that the measure of student's IQ does not seem highly interrelated with any other of the variables, save for a moderate-to-low

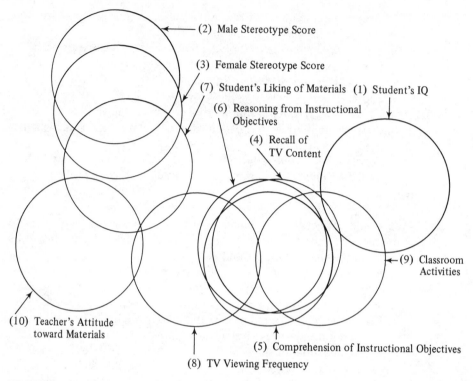

(2) Male Stereotype Score

(3) Female Stereotype Score

(7) Student's Liking of Materials (1) Student's IQ

(6) Reasoning from Instructional
    Objectives

(4) Recall of
    TV Content

(9) Classroom
    Activities

(10) Teacher's Attitude
    toward Materials

(5) Comprehension of Instructional Objectives

(8) TV Viewing Frequency

FIGURE 13.1   Measurement Overlap of all correlations greater than .45 (with some distortions to facilitate illustration)

correlation with classroom activity frequency. Finally, it is evident that some of the "causal" variables are interrelated with "effect" variables, such as TV viewing frequency with the comprehension measures.

These clusters of measurement overlap do tend to show us more of a basic, underlying structure than do the ten individual variables. We could thus reduce our description of this study from ten variables to fewer variables and have a more basic and more economical way to characterize the behavior we are studying.

What we have done in plotting the relations among the various measures as in Figure 13.1 is essentially what the mathematical technique of factor analysis provides. Put into practical terms, factor analysis takes a matrix of intercorrelations such as shown in Table 13.1 and provides a method for determining, on the basis of the correlations among a relatively large number of variables, what fewer, more basic, and unique variables might underlie this larger number of variables. A factor-analysis solution was conducted of these intercorrelations, and the results are summarized in Table 13.2.

Table 13.1   Intercorrelation Matrix for Variables in the Simulated Study of Use of Television to Reduce Sex Stereotyping

Variables	Correlation with Each Other (Variables Designated by Number)								
	(2)	(3)	(4)	(5)	(6)	(7)	(8)	(9)	(10)
(1) Student's IQ	.07	.07	.31	.16	.20	−.07	−.01	.45	−.04
(2) Male stereotype score		−.89	.06	.34	.08	−.37	.37	−.15	.30
(3) Female stereotype score			−.11	.43	.09	−.57	.30	−.07	.21
(4) Recall of TV content				.89	.97	.10	.68	.83	−.03
(5) Comprehension of instructional objectives					.83	−.10	.67	.70	.00
(6) Reasoning from instructional objectives						.11	.67	.70	.00
(7) Student's liking of materials							.23	−.08	.46
(8) TV viewing frequency								.19	.19
(9) Classroom activity frequency									−.22
(10) Teacher's attitude toward materials									

Table 13.2   Factor Matrix (Varimax Rotation) for the Intercorrelations Shown in Table 13.1 and Figure 13.1

Variables	Factors		
	I	II	III
(1) Student's IQ	.31	.12	−.29
(2) Male stereotype score	−.06	−.90	−.21
(3) Female stereotype score	−.09	−.97	−.03
(4) Recall of TV content	.99	−.01	.01
(5) Comprehension of objectives	.88	−.34	.01
(6) Reasoning from objectives	.97	−.02	.07
(7) Student's liking of materials	.10	.60	.69
(8) TV viewing frequency	.68	−.27	.52
(9) Classroom activity frequency	.84	.16	−.42
(10) Teacher's attitude toward materials	−.04	−.13	.66

Notice first that the factor matrix (Table 13.2) lists the factors as Columns I, II, and III. This indicates, so to speak, that three dimensions of common variance were found in the analysis (similar to the three clusters portrayed in Fig. 13.1). The numbers given within the matrix itself are interpreted as correlation coefficients. They tell us the correlations between each of the original measurements and each of the three factors, or hypothetical underlying variables found in the analysis. Thus we can see, as was the case in Figure 13.1, that the measures of recall, comprehension, and reasoning from program content all tend to be highly correlated with what is defined as Factor I, whereas the two attitudinal measures of stereotyping tend to be highly correlated with Factor II. When we consider a variable's correlation with a factor, we talk about this correlation in terms of *factor loading.* That is, we would say that the measures of recall, comprehension, and reasoning from objectives are all "highly loaded" on Factor I, and that the stereotype measures "load" on Factor II.

Results and Interpretations

A *factor matrix* (or, as discussed later, a *rotated matrix*) is mainly what the mathematics of factor analysis provides us. The interpretation of what each factor means, or how we might label it, is left to the subjective evaluation of the researcher. Usually this evaluation involves close examination of the factor matrix and asking such questions as: What variables tend to load highest on each factor? How might these variables together define a more basic variable relevant to what is being studied? In interpreting the factor matrix shown in Table 13.2, we might report the following conclusions:

> Three factors were extracted in the analysis. These were labeled as follows: (I) *comprehension/exposure,* defined by dependent variable measures of recall, comprehension, and reasoning, and independent variables of exposure to the TV program and classroom activities; (II) *stereotyping/student's liking of materials,* defined by the inverse relationship of the degree to which students would impose sex-role stereotypes upon occupations and the degree to which they indicated liking of the program and activity materials; and (III) *teacher's attitude/student's liking of materials,* defined as the simple relationship between those two particular variables. For subsequent analyses it appears useful to combine the dependent-type variables of recall, comprehension, and reasoning into a single underlying variable designated as "understanding." It also appears useful to combine the separate measure of male and female stereotyping into one overall "stereotyping" score.

Another point concerning the interpretation of the factor matrix is that factor loadings (as in Table 13.2), like correlation coefficients, have signs. Usually, unless a minus sign appears, the relationship between the variable and the factor is assumed to be positive. That is, the plus signs are simply omitted. Such signs, as was the case with correlation coefficients, are indicative of the direction of the relation. Thus, the more that the students indicated a liking of the materials, the less they tended to impose sex-role stereotypes in their assignments of people to occupations (see Factor II). The variables with high negative loadings are just as pertinent to factor interpretations as are the variables with high positive loadings. Loadings have different signs on the given factor mean that the variables' scores have different directions of relationship.

Probably the most frequent use of factor analysis in communication and education research is where the researcher wishes to see if a relatively large number of measures can be reduced to fewer, more basic, underlying variables, as in how items on a test of some type may represent fewer variables than the items themselves, as was the case with the measures of recall, comprehension, and reasoning, or the stereotyping scores, in the foregoing example. If you have taken achievement examinations, you probably know that the test items on such exams reduce to factors such as mathematical or verbal reasoning ability. A frequent use of factor analysis is similar to the present example where there is a desire to see what type of structure underlies the variables in a study. Parts of a factor structure may simply represent redundancy among certain measures or relationships of causality in others. We have chosen this latter example because we shall use parts of it in subsequent chapters to show relationships among different multivariate approaches.

THE LOGIC OF FACTOR ANALYSIS

Although the detailed mathematical logic of factor analysis is too complex a topic to present here, we can provide a rather general view of what a factor is, and further, how we can rotate factor structures in order to provide the best fit, or most useful fit, of the factors to the data, or to whatever theoretical structure is being considered.

Factors and Rotation

As we said earlier, factors are a redefinition of the interrelations among variables according to how these interrelations might be defined in terms of fewer (hypothetical) variables, In order to illustrate what we mean by this, let us consider a simplified problem involving five variables and eventually two factors.

Consider first the matrix of intercorrelations presented in Table 13.3 (A). The correlation coefficients define for us the interrelations among the variables A, . . . E. Simply by inspection of this table we can get the hunch that variables A and B tend to vary together more

Table 13.3   Simplified Example of (A) Intercorrelation Matrix, (B) Unrotated Factor Matrix, and (C) Rotated Factor Matrix

(A) *Intercorrelations*

		Variables			
		B	C	D	E
	A	.985	.500	.342	.174
Variables:	B		.642	.500	.342
	C			.985	.940
	D				.985

(B) *Unrotated Factor Matrix*

		Factors	
		I	II
	A	.574	.819
	B	.423	.906
Variables:	C	−.423	.906
	D	−.574	.819
	E	−.707	.707

(C) *Rotated Factor Matrix*

		Factors	
		I	II
	A	.996	.087
	B	.966	.259
Variables:	C	.423	.906
	D	.259	.966
	E	.087	.996

so than either with variables C, D, and E, and that C, D, and E seem to go together. We could plot these interrelations in terms of overlapping circles (as in Fig. 13.1) if we wanted to. For present purposes, however, a better way to plot these intercorrelations is shown in Figure 13.2 (A). Each arrow (vector) here represents one of the variables. The angle between any two variables reflects the correlation between them. The smaller the angle between any two vectors, the greater is their correlation. (In order to provide this illustration we have taken the cosine of each angle between two variables as equal to its respective correlation coefficient. For example, the angle between A and B is 10° and its cosine is .985, the correlation between A and B; or the angle between A and E is 80° and its cosine is .174, the correlation between A and E. This has nothing to do with the logic of factor analysis—it is simply necessary to provide this illustration.)

What we have, then, in Figure 13.2 (A) portrays only the interrelations among the five variables. By measuring the angles between all possible pairs of vectors, we could provide the correlation coefficients shown in Table 13.3 (A). These coefficients are what we begin with in conducting a factor analysis.

Suppose, next, that on the basis of certain mathematical criteria

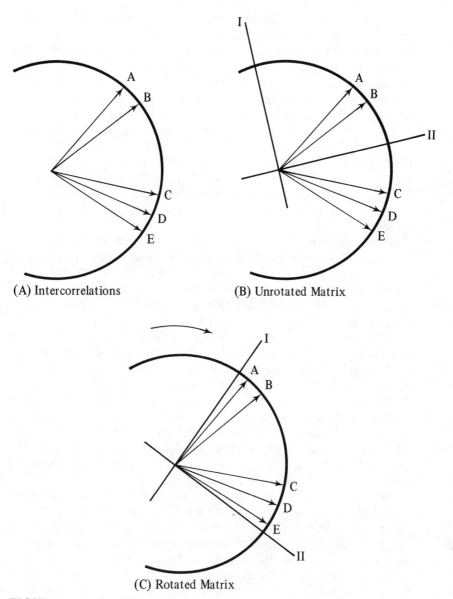

(A) Intercorrelations   (B) Unrotated Matrix

(C) Rotated Matrix

FIGURE 13.2 Graphic portrayal of (A) intercorrelations, (B) unrotated matrix, and (C) rotated matrix for data given in Table 13.3. The correlation between any two vectors is equal to the cosine of the angle between them. Variables are vectors A, . . ., E, factors are vectors I and II.

(which differ for various types of solutions), we considered the relations among the five variables not in terms of interrelations among each other, but all in relation to two factors (I and II). For purposes of illustration, suppose that these two factors were initially defined as shown in Figure 13.2 (B). A first point to consider is that the correlations between the variables and the two factors are reflected in the angles (actually cosines of the angles) between each of the vectors representing variables and the two vectors now representing factors. We can still define the relations among the variables, but now we do it in terms of the two factors as reference points. In other words, we can define the relations among the variables in terms of how they correlate with the two factors. These correlations, or factor loadings, are presented in Table 13.3 (B), called an *unrotated factor matrix*. Another thing to consider at this point is that the two factors are treated as completely uncorrelated with each other (that is, 90° apart in Fig. 13.2 (B), and the cosine of 90 equals 0, their correlation). This is a common assumption in many types of factor-analytic solutions, but is not a necessary assumption by any means.[2]

If we want these two factors to reflect as clearly as possible what we might consider to be a hypothetical variable underlying variables A and B, and another variable underlying variables C, D, and E, we rotate their position (keeping them 90° apart) as shown in Figure 13.2 (C). Notice now that we have made the best compromise possible in locating Factor I with as small angles (or as large a correlation) as possible in locating it near variables A and B, while at the same time positioning Factor II as closely as possible to variables C, D, and E. When we now consider the correlations of the five variables with the two factors, the pattern is relatively clear-cut. In Table 13.3 (C), notice that variables A and B have relatively high loadings on Factor I (.996 and .966, respectively), while having relatively lower loadings on Factor II (.087 and .259, respectively). Similarly, variables C, D, and E all have relatively high loadings on Factor II (.906, .966, and .996, respectively), but relatively low loadings on Factor I (.423, .259, and .087, respectively).

The rotation, then, has provided a readily interpretable factor structure, as can be seen by assessing the factor loadings in the rotated factor matrix in Table 13.3 (C). We would probably conclude that two more basic variables, as defined by Factors I and II, underlie the five measurements represented in variables A, . . . , E. In an actual case of using factor analysis, we would use this rotated factor matrix as a basis for reporting our conclusions (as we did with reference to Table 13.3).

A word of warning is in order, however. What we have seen in the foregoing example is a highly simplified illustration of factors and of factor rotation. For one thing, the illustration said nothing about measurement error. That is, presumably some of the variance in a

---

[2] When factors are treated independently of one another, this is called an orthogonal solution, whereas consideration of factors as interrelated is called an oblique solution.

collection of measurements is due to error and thus must be some-how defined in the factor-analytic solution. Neither did the illustra-tion accommodate the fact that variables typically have some portion of their variance that is specific to them alone, and may not be in-corporated in the factor structure. Additionally, the preceding illus-tration gives no hint of the computational procedures involved in factor analysis. There are mathematical criteria (as well as some subjective decisions necessary to use these criteria) involved in de-termining the number of factors to be extracted, in locating these factors prior to rotation, and in rotating the factor structure. Even if we were to undertake a computational solution of the present simple problem, it would take considerable time for hand calculation. Evi-dence of the computational complexity of factor-analytic solutions is that many of our modern methods would not be economically fea-sible (or even possible in many cases) were it not for the devel-opment of electronic computers. Finally, factor-analytic solutions of-ten involve extraction of more than two factors. Obviously, we would be hard pressed to present graphically more than two factors at a time when using illustrations like Figure 13.2; the present illus-tration is a simple one indeed. At best, all we have done is to pre-sent a rudimentary picture of the relations between a correlational matrix and unrotated and rotated factor matrixes.

Relation Between the Correlation Matrix and the Factor Matrix

Another way of viewing the essence of factor analysis is to consider the relationship between what is defined in the correlation matrix and what is defined in the factor matrix. Recall that we have charac-terized factor analysis as the redefinition of the interrelations among variables in terms of relations of these variables with a set of (fewer) hypothetical variables. If we consider that both the correlation and factor matrixes are two different ways of defining the same inter-relations among a set of variables, then we can define the statistical goal of factor analysis as the substitution of a factor matrix for a correlation matrix. We can get some idea of the relation between these matrixes by comparing tables (A), (B), and (C) in Table 13.3. (Again, bear in mind that the data here are fictitious and incorporate no measurement error.) We can state a relation as follows: The cor-relation between any two variables as shown in the correlation ma-trix (for instance, r_{AB}) is equal to the sum of the products of the two variables' loadings on each factor, as shown in any form, unrotated or rotated, of the factor matrix [for example, $(A_IB_I) + (A_{II}B_{II})$].[3] No doubt this statement seems awkward and confusing at first. (If we

---

[3] Where A_I means the loading of variable A on Factor I; B_{II} the loading of variable B on Factor II; and so on.

took the time to define the necessary mathematical notations, we could state it much more parsimoniously.) What it simply says, for example, is this: If we take the product of the loadings of variables A and B on Factor I say, $.574 \cdot .423$ in the unrotated matrix in Fig. 13.2 (B) and add it to the product of the loading of these same two variables on Factor II ($.819 \cdot .906$), then this sum will equal the correlation between variables A and B ($r_{AB} = .985$ in the correlation matrix). The same holds true between our rotated matrix and the correlation matrix — that is,

$$(A_I B_I) + (A_{II} B_{II}) = r_{AB}$$
$$(.996 \cdot .966) + (.087 \cdot .259) = .985$$

For another example, consider the correlation between variables C and E ($r_{CE} = .940$).

$$(C_I E_I) + (C_{II} E_{II}) = r_{CE}$$
Unrotated: $(-.423 \cdot -.707) + (.906 \cdot .707) = .940$
Rotated: $(.423 \cdot .087) + (.906 \cdot .996) = .940$

Another way, then, of viewing the logic of factor analysis is simply that of matrix substitution, or what we have called redefinition.

Rotational Schemes

In the present example, we rotated the two factors so that we could get maximum loading of variables A and B on Factor I, while getting maximal loading of variables C, D, and E on Factor II. Or we could look at this in terms of getting A and B high on Factor I, but getting C, D, and E low on this factor, or getting C, D, and E high on Factor II, but A and B low on this factor. This kind of a rotation comes close to illustrating what we call rotation to *simple structure*. There are, however, a number of different procedures for factor rotation. Suppose, by contrast, that our theory led to the prediction that there should be two factors, and that certain variables would load on each. In this case we would rotate the axes to where theory suggests that they fit; this is called rotation to *psychological fit*.

One facet of research on factor-analytic methods has centered on development of rotational schemes that are as objective as possible. Two of these schemes that attempt to objectify mathematically the criterion of simple structure are the *quartimax* and *varimax* methods.

Sources of Variation

Earlier we mentioned that our simplified example said nothing about measurement error, or how we accommodate nonerror variance that is not incorporated in a factor structure. We can think of both of these aspects as potential sources of variation to be treated in a factor analysis. For present purposes, let us label them as:

V_e = error variance
V_s = nonerror variance specific to variables and not included in the factor structure

In addition to the above sources of variation, we have that variance which is incorporated into the factors defined in the analysis; we call this, *common factor* variance, or V_c. Common factor variance, or V_c, is in turn composed of the variance attributed to *each* of the factors determined in the analysis – that is,

$$V_c = V_I + V_{II} + V_{etc.}$$

We consider that all of the foregoing sources of variation together, define what we call *total* variance or V_t. Thus, for a given analysis, we can conceive of a total amount of variance (V_t) being divided into the various sources of variation. In a three-factor structure, for example.

$$V_t = V_I + V_{II} + V_{III} + V_s + V_e$$

Table 13.4 presents a complete matrix for a hypothetical solution involving five variables (M, . . . , Q) and where three factors (I, II, III) have been extracted. With five variables, the total variance (V_t) would equal 5.0. By examining this matrix we can see the division of this total variance into its various sources. In terms of actual values of variance, we can see how they are divided among the factors (V_I, V_{II}, V_{III}) as well as according to specific variance (V_s) and error variance (V_e), thus:

Table 13.4   Example of a Rotated Factor Matrix Including All Sources of Variation

		Factors			Sources of Variation		
		I	II	III	(V_c)	V_s	V_e
	M	.10	.10	.90	(.83)	.07	.10
	N	.08	.65	.65	(.85)	.08	.07
Variables:	O	.95	.06	.04	(.91)	.06	.03
	P	.90	.20	.06	(.85)	.01	.14
	Q	.08	.88	.02	(.78)	.15	.07
Variance:		1.74	1.25	1.23	(4.22)	0.37	0.41
Proportion of total variance:		.35	.25	.25	(.85)	.07	.08
Proportion of common variance:		.41	.30	.29			

$$V_t = V_I + V_{II} + V_{III} + V_s + V_c$$
$$5.0 = 1.74 + 1.25 + 1.23 + 0.37 + 0.41$$

It is usually more convenient, however, to view these sources of variation in terms of *proportion of total variance;* thus:

$$V_t = V_I + V_{II} + V_{III} + V_s + V_e$$
$$1.0 = .35 + .25 + .25 + .07 + .08$$

Given the above overall view, the interest then centers on what has been defined as common factor variance (V_c). In Table 13.4, V_c is equal to 4.22 in terms of actual variance, and this is approximately .85 of the total variance. In focusing on common factor variance, it is further useful to see how the individual factors each account for this variance; this is called *proportion of common factor variance* (as contrasted with the proportion of total variance considered above for each factor). In terms of such proportions, we would say that:

$$V_c = V_I + V_{II} + V_{III}$$
$$1.0 = .41 + .30 + .29$$

What we have described thus far in Table 13.4 has centered only on the overall sources of variation. We can next examine the matrix in more detail and see how the individual variables (M, . . ., Q) reflect sources of variation. Recall that factor loadings are the correlations of each variable with a factor. Like correlation coefficients, these loadings when squared tell us the variance that the two variables have in common. For example, if the loading of variable M on Factor I is .10, then we can say that this contributes .01 (that is, .10 · .10) as a source of variation. Notice in Table 13.4 that if we take the three factor loadings for variable M (.10, .10, .90) and square them (.01, .01, .81) we have the amount of common variance (.83). or *commonality*, attributed to that variable. In other words, on the basis of the factor loadings shown in the matrix, we can break down the sources of common variance according to the relations between the variables and the factors.

It might be worthwhile to digress for a moment, and to consider how the foregoing breakdown would apply to the earlier sample rotated factor matrix shown in Table 13.3. Remember, we noted earlier that neither specific variance nor error were considered in this example. Indeed, what we actually assumed was that there was no error nor specific variance. In this case the sum of the squared factor loadings on each variable should equal 1.0, since the total variance in this matrix is all common variance. This, indeed, is the case: for example, taking the loadings of variable A on Factors I and II in the rotated matrix (Table 13.3), we have .996 and .087. When squared (and rounded) we have .99 and .01, which when added are equal to 1.0.

Returning to Table 13.4, we can also see how the individual factor loadings can be seen in relation to the variance attributable to each factor extracted in the analysis. This time we square the factor loadings and sum *down* the column. Taking Factor I in Table 13.4 as an example, we take the loading of each variable (.10, .08, .95, .90, .08), square each (.01, .0064, .9025, .81, .0064), then sum, and we have the amount of variance attributable to this factor (1.74). In analyses such as this, the sum of squared factor loadings is called an *eigenvalue*.

What we have seen, then, in all of the foregoing is how a factor-analytic solution can be interpreted in terms of the various sources of variation. We can consider the overall picture in terms of variance attributable to common variance (or the individual factors), to specific variance, and to error variance. Or we can consider variance attributable to each factor in terms of its relative proportion of common variance. In more detail, we can consider how the individual variables contribute as sources of variance. In this case, the contribution of each variable is taken as the square of its loading on each factor.

SOME PRACTICAL ASPECTS OF FACTOR ANALYSIS

Given a set of measurements on a number of variables, what are some of the practical considerations reflected when factor analysis is employed? Because we have not gone into any detail in terms of computational procedures, we can do no better than to discuss some of the high points of these considerations.

Measurement Error

Given a matrix of intercorrelations among the variables to be factor analyzed, one problem concerns defining the error variance (V_e) attributable to each variable. Suppose that, on the basis of prior research, we knew how much nonerror variance was attributable to a given variable. We could describe this nonerror variance on the basis of determining the correlation of the variable with itself in some type of test, retest situation. The square of this correlation coefficient would then be what we would consider as nonerror variance. Examine the matrix shown in Table 13.5. This is the same correlation matrix as shown in Table 13.3, except that this time we have entered a "1.0" where each variable intersects itself. These entries indicate that each variable is perfectly correlated with itself— there is no error variance (as was assumed in the example where

Table 13.5   **Example of the Intercorrelation Matrix from Table 13.3, but Where Assumption of No Error Variance Is Indicated by Including "1.0" Entries in the Diagonals**

	A	B	C	D	E
A	(1.0)	.985	.500	.342	.174
B		(1.0)	.642	.500	.342
C			(1.0)	.985	.940
D				(1.0)	.985
E					(1.0)

this matrix was factored). If we made this assumption in running a factor analysis, we would incorporate such entries in the correlation matrix used as input for the analysis. (This is what is meant by "putting ones in the diagonals") On the other hand, if we have some basis for estimating what this entry might be, some figure of less than 1.0 is located in the matrix for each variable. As mentioned earlier, such entries might be based on prior research. There are, however, a number of mathematical procedures for estimating these entries on the basis of characteristics of the variable's measurement distribution.

Given, then, assumptions about the measurement error attributable to each variable in an analysis, it remains for the solution itself to allocate the remaining variance on each variable (that is, $V_t - V_e$) in terms of that which is attributable to factors (V_c) and that which remains as specific to the variables (V_s).

Solution Model

Without getting into computational details, it is difficult to say much about the different types of factor-analytic solutions. When we consider a particular solution model, this is a decision somewhat in addition to how we have defined measurement error and what rotational procedures will be undertaken. We say somewhat, because some solutions are based on certain assumptions concerning measurement error, and some solutions carry particular implications concerning rotation, or even the need for rotation. We say in addition, because the solution itself centers on how factors are basically extracted from the correlation matrix, whereas the estimates of measurement error and rotational procedures involve additional decisions. Perhaps the most frequently encountered solution is one known as the *principal components* method. Suffice to say that the key feature of this method is that it attempts to extract a maximum amount of variance as each factor is calculated; it attempts to redefine the correlational matrix with as few factors as possible. Oftentimes, one will encounter this method, combined with an assumption of no measurement error (putting ones in the diagonals), and with use of the varimax rotational method.

Interpretation of the Rotated Factor Matrix

A practical point is to reemphasize that the interpretation of factors is a subjective decision on the part of the researcher. Nothing in the logic of a factor-analytic solution tells a researcher how to label a particular factor. Interpretation of a factor involves examining what variables appear to load relatively high on that factor (and perhaps low on all others), then asking oneself if there is some more basic and meaningful variable that may underlie these high loading variables. If this more basic variable can be given a meaningful name, we use this name to label the factor. Thus, in the example given early in the chapter, we said that measures of recall, comprehension, and reasoning all seemed to index a basic variable that we might call "understanding."

Given a particular solution where the factors have been meaningfully labeled, it is often important to make an overall evaluation of how the factors together characterize some overall meaningful construct. For example, we said earlier that all of the factors (Table 13.2) seemed to characterize the basic variables relevant to the use of television and classroom activity to reduce sex-role stereotyping. We can then consider how each of the facets of this construct variously contributes to the overall picture in terms of how common variation (V_c) breaks down into variation due to each of the factors (that is, V_I, V_{II}, and so on). Finally, we can see how the construct itself fits into the totality of what we have measured. That is, we can consider how much of the total variation (V_t) is accommodated by the factor structure itself (V_c), and how much variation remains specific to the original variables (V_s) and to measurement error (V_e).

Factor Scores

Often when a researcher has a further analysis that follows the results of a factor analysis, *factor scores* are used. Such scores are simply a combination of variables that loaded on a factor. Thus, for example, suppose in the sample study in this chapter that we wished to compare boys and girls in terms of their overall understanding of the television materials. Recall that the factor analysis yielded a factor (I) that reflected primarily the combination of the individual measures of recall of television content (variable 4), comprehension of objectives (5), and reasoning from objectives (6). In this case we might add or average these three variables together to form one composite variable for the comparison of boys' and girls' understanding. This combination could also be calculated by taking the variable loadings as a basis for "weighting" their contribution to the overall score. Such combinations, which may take a variety of forms, are generally known as *factor scores*. Many computer programs provide options for calculating factor scores as the weighted combination of all variables loading on a factor.

SUMMARY

From a practical standpoint, *factor analysis* is a method by which the relations among a relatively large number of variables are redefined in terms of relations with relatively fewer hypothetical variables called *factors*. Factor-analytic procedure begins with a matrix of intercorrelations and estimates of the measurement error involved in each variable. The solution itself redefines the correlational matrix in terms of a *factor matrix*. There are then different procedures for *rotating* the factor matrix according to such criteria as *simple structure or psychological fit*. Given a rotated factor matrix, the researcher then attempts a subjective interpretation of the factors that have been obtained. Sources of variation involved in a factor-analytic solution include definition of *common factor variance* (variance attributable to the factors extracted), *specific variance* (nonerror variance not included in the factor structure), and *error variance*.

SUPPLEMENTARY READINGS

Harman, Harry H., *Modern Factor Analysis*. Chicago: University of Chicago Press, 1967. Still the classic, but technical.

Kerlinger, Fred N., *Foundations of Behavioral Research,* 2nd ed. New York: Holt, Rinehart and Winston, 1973. Chapter 37 is a practical example of the use of factor analysis. Not too technical for a reader with a little background in statistics.

Kerlinger, Fred N., and Elazar J. Pedhazur, *Multiple Regression in Behavioral Research.* New York: Holt, Rinehart and Winston, 1973. Chapter 13 is a brief overview of factor analysis, especially as related to multiple regression and multivariate analysis of variance.

Monge, Peter, and J. Cappella (Eds.), *Multivariate Techniques in Communication Research.* New York: Academic Press, 1978. An excellent review of all multivariate techniques; somewhat technical.

Nie, Norman H., C. Handlai Hull, Jean G. Jenkins, Karin Steinbrenner, and Dale H. Bent, *Statistical Packages for the Social Sciences,* 2nd ed. New York: McGraw-Hill, 1975. Chapter 24 is clear and only so technical as one needs to set up and interpret a packaged computer program for factor analysis.

CANONICAL CORRELATION

14

In the previous chapter we saw how factor analysis could be used to identify patterns of relationship among and within groups of individual variables. Sometimes we are interested in describing the relationship between two clusters of variables. In this case we employ a statistical method known as canonical correlation.

THE NATURE OF CANONICAL CORRELATION

A canonical correlation analysis between two sets of variables yields one or several canonical correlation coefficients (R_c, or R_c^1, R_c^2 and so on) that index the degree of correlation between *the two sets* rather than the individual variables. Somewhat akin to linear regression analysis, the canonical analysis seeks the maximum linear combination between linear combinations for each set of variables. Each linear combination is called a *canonical variate*. The relationship between a pair of variates yields a canonical correlation coefficient (interpreted like other correlation coefficients; see Chapter 10). Each variate has a set of weights indicating the relative relation of each variable in the set to the variate. Depending on the complexity of the relations between the two sets of variables, different successive linear combinations will yield successive pairs of canonical variates, the associated canonical correlation coefficients, and variable weightings.

A SAMPLE STUDY

Problem

Recall that in Chapter 13 we introduced a simulated study involving the use of a television series and classroom activities to reduce sex-role stereotyping in children's considerations of different occupations. (The variables in this study were summarized on pages 162–163, the intercorrelations among them in Table 13.3 (page 165), and the results of a factor analysis in Table 13.2 (page 165). In the last chapter we saw how the variables in the study tended to cluster into three factors:

    I. Comprehension/exposure
   II. Stereotyping/students' liking of materials
  III. Teacher attitude/students' liking of materials

One set of relationships of particular interest to us among these variables is the degree to which measures relevant to students' experiences with the materials can be related to their reactions to them. Or, put more practically: To what degree can we predict the multiple effects of the TV program and activities, given data on experience with them?

Method

To begin, we could divide the variables into two clusters representing independent and dependent variables:

Independent variables (predictors)
  (8) TV viewing frequency
  (9) Classroom activity frequency
(10) Teachers' attitudes toward materials
Dependent variables (to be predicted)
  (7) Students' liking of materials
(4, 5, 6) Students' overall understanding of the materials, based on combinations of recall, comprehension, and reasoning measures
  (2, 3) Stereotyping score, based on a combination of measures of the degree to which children use male or female stereotypes in assigning individuals to occupational categories

The foregoing lists incorporate several modifications from the original list of variables presented in Chapter 13 (pages 162–163). For one thing, we have omitted students' IQ (variable 1) because it did not show any significant relationship with any variables of interest to us in the present analysis. Also, the factor analysis in the prior chapter (Table 13.2) indicated that the formerly separate measures of recall, comprehension, and reasoning were all so substantially interrelated that they could just as well be combined into one overall variable of "understanding." The factor analysis further indicated that there was such a high degree of relationship between male and female stereotyping measures that the two were combined into an overall stereotyping measure. These are practical examples of the utility of conducting a factor analysis as a basis for combining redundant variables prior to entering into subsequent analyses. This simplifies our canonical correlation analysis; however, the analysis could be done with the original variables of understanding and stereotyping, rather than the combined ones.

Again it is useful for us to consider the correlations among the variables just listed, and we can do so in terms of a matrix of intercorrelations such as that given in Table 14.1. From this table it is not difficult to estimate that the overall understanding score (variables 4, 5, 6) is most related to television viewing (variable 8) and classroom activity frequencies (variable 9), whereas the stereotyping measure (variables 2, 3) seems most related to the teacher's attitudes (variable 10) toward the materials. Any canonical correlations calculated to show the relationship between this cluster of independent and dependent variables should reflect these patterns of relationship. Indeed, the canonical correlation gives us a way of stating this relationship more concisely by attaching correlational indexes directly to it.

Results and Interpretations

Table 14.2 summarizes the results of a canonical correlation analysis of the intercorrelations among the variables shown in Table 14.1. Again, we should bear in mind that as a part of setting up the analysis we initially differentiated two clusters of variables which we referred to as "independent" and "dependent" in this particular study. The analysis (Table 14.2) indicates that two configurations of correlation (two pairs of canonical variates) were obtained in the analysis. Both indicate a relatively high correlation between a configuration of independent and dependent variables. We can interpret those patterns by examining the relative weightings of each variable in the two stages of the analyses.

In the first canonical correlation, it is quite evident that the pattern is one where frequency of viewing television and the frequency of classroom activities are good joint predictors of a child's overall

Table 14.1   Intercorrelations Among Variables in the Sample Study of Sex-Role Stereotyping

Variables	Variables (By Number from Left Column)				
	(9)	(10)	(7)	(4, 5, 6)	(2, 3)
(Independent)					
(8) TV viewing frequency	.19	.19	.23	.72	−.35
(9) Classroom activity frequency		−.22	−.07	.80	.11
(10) Teachers' attitudes			.46	−.02	−.27
(Dependent)					
(7) Students' liking of materials				.05	.48
(4, 5, 6) Overall understanding*					−.18
(2, 3) Stereotyping score*					

*Combined from variables in Table 13.1.

Table 14.2   **Results of a Canonical Correlation Analysis of Intercorrelations Shown in Table 14.1**

First canonical correlation, $R_c1 = .97$
Variable weights:

Independent			Dependent		
(8)	TV frequency	.59	(4, 5, 6)	Overall understanding	1.00
(9)	Activity frequency	.71	(2, 3)	Sex-role stereotyping	.03
(10)	Teacher attitude	.03	(7)	Liking of materials	.03

Second canonical correlation, $R_c2 = .84$

Variable Weights:

Independent			Dependent		
(8)	TV frequency	.43	(4, 5, 6)	Overall understanding	−.23
(9)	Activity frequency	−.37	(2, 3)	Sex-role stereotyping	−.98
(10)	Teacher attitude	.71	(7)	Liking of materials	1.02

understanding of the materials in the package ($R_c1 = .97$). We can also note that teacher attitude seemed to have little relationship to overall understanding of the materials. And by the same token, the frequency of the television programs or classroom activities seemed to have little relationship to the degree of sex-role stereotyping, or to a student's liking of the materials.

The second canonical correlation for the same two clusters indicates that beyond the relationship just noted, there is the further relationship that the teacher's attitude toward the materials tends to predict the student's liking of the materials, but also predicts that a student will engage in less sex-role stereotyping ($R_c2 = .84$). We can also see that to a lesser degree, the more television programs a child watches and, to a minor extent, the fewer activities engaged in, also make this same prediction.

Stated even more concisely, the researcher could report the following conclusions:

> Results of the canonical correlation analysis indicated a high degree of relationship of television program and classroom activity frequencies with the child's overall understanding of the materials. The same analysis also indicated a high, but somewhat lesser, relationship between the teacher's attitude toward the materials and a student's liking of the materials, as well as a reduction in sex-role stereotyping when students consider individuals for occupations.

FURTHER NOTES ON CANONICAL CORRELATION

Significance

It is possible to test the statistical significance of canonical variates relative to the degree that linear relations between the two variable sets have been extracted. This involves the calculation of what is called Wilk's *lambda* and its test of significance by a chi-square method. Most references on canonical correlation also include a discussion of Wilk's *lambda*.

Relation with Factor Analysis

As we mentioned at the outset of Chapter 13, multivariate analyses are interrelated in the sense that they all involve different ways of characterizing measurement overlap (or common variance) among or within clusters of variables. It is not difficult, for example, to see relationships between the results of the canonical correlation analysis shown in Table 14.2 and the factor-analysis results summarized in Table 13.2. Recall how television program and classroom activity frequencies tended to load on Factor I in that analysis along with the various measures of recall, comprehension, and reasoning, the latter being collapsed for this analysis into one overall variable of understanding. It can be noted too in the results of the factor analysis (again Table 13.2) that the variables of teacher attitude, the student's liking of the materials, and the two original measures of sex-role stereotyping tend to appear on common factors. At the same time, however, it should be noted that the canonical correlation results are not an exact reflection of the factor-analysis results. This is because the canonical correlation analysis attempts to maximize variance explained between two clusters of variables whereas a factor analysis tries to reduce the variance in an overall matrix to relatively fewer dimensions. Put into more practical terms, we can expect canonical correlations to produce structures that are biased toward maximizing the relationship between the two clusters of variables or the prediction of one cluster from another, and this will vary somewhat, although not totally, from the results of a factor analysis of the same variables.

It is further useful to bear in mind the similarity between the two analyses. Roughly speaking, canonical variates are like "factors." A squared canonical correlation coefficient, like the sum of squared loadings on a factor, is called an *eigenvalue*. Finally, the variable weightings on a canonical variate are interpreted loosely akin to factor loadings.

Relation with Regression Analysis

Just as we can envisage the logical relation between a correlation and a regression analysis, or a multiple correlation and a multiple regression analysis, the results of a canonical correlation analysis can be used to predict the values of one cluster of variables given the values of variables in the other. Although this approach is beyond the scope of this text, the basic logic of canonical correlation analysis underlies what is called *multivariate multiple regression analysis.* As implied by the term *multivariate,* this type of regression approach provides for more than one dependent variable in an analysis.

Relation with Multivariate Analysis of Variance

Recall from Chapters 10, 11, and 12 how values of "0" and "1" could be used to code group differences as a variable in a correlation or regression analysis. These were called "dummy" or "group" variables. Examples were also given in Chapters 11 and 12 to show the relationship of a regression analysis incorporating dummy variables with an analysis of variance where the dummy variables had been independent variables in the form of groups or treatments. This same comparison can be carried over to a canonical correlational analysis where dummy variables can be included among independent variables. This provides a basis for assessing group or treatment effects on multiple dependent variables that is essentially accomplished in what is called *multivariate analysis of variance.*

SUMMARY

Canonical correlation describes the correlation of one cluster of variables with another. Its basis is the calculation of the maximum linear combinations between two sets of variables. Each combination is a pair of *canonical variates* that yield a *canonical correlation coefficient* and a set of coefficients that describe the weighting of each variable in the two sets on the canonical variates. Successive variates attempt new maximum linear correlations; hence for a particular analysis there may be more than one correlation coefficient and sets of variable weights.

Canonical correlation has much in common with factor analysis and multiple regression analysis. Categorical variables may be entered into the analysis by use of binary or dummy variables.

SUPPLEMENTARY READINGS

Cooley, W. W., and P. R. Lohnes, *Multivariate Procedures for the Behavioral Sciences.* New York: Wiley, 1962. Canonical correlation is presented in Chapter 3.

Kerlinger, Fred N., and E. J. Pedhazur, *Multiple Regression in Behavioral Research.* New York: Holt, Rinehart and Winston, 1973. Chapter 12 presents a practical overview of canonical correlation with an emphasis on its relation to multiple regression.

Monge, Peter, and J. Cappella (Eds.), *Multivariate Techniques in Communication Research.* New York: Academic Press, 1978.

Nie, Norman H., C. Handlai Hull, Jean G. Jenkins, Karin Steinbrenner, and Dale H. Bent, *Statistical Packages for the Social Sciences.* New York: McGraw-Hill, 1975. Chapter 25 describes practical use of a computer program for canonical correlation.

MULTIPLE DISCRIMINANT ANALYSIS

We saw in Chapters 11 and 12 on regression analysis how it is possible to use one or more variables, including binary or dummy variables (group or treatments coded as "0" or "1"), to predict values of a variable. By contrast, when we use two or more variables to predict membership in categories, groups, or any such categorical variable, the method known as multiple discriminant analysis is used. Applications of multiple discriminant analysis are usually one of two types: One is to see the degree to which members and different groups (for example, successful teachers as against unsuccessful ones) can be differentiated in terms of an array of discriminator variables (amount of training, time in service, size of school system, and so on). Another use is to classify new individuals or phenomena into categories, given the results of a previous analysis of the relationship between discriminating variables and categories (for example, predicting whether teachers are likely to be successful or unsuccessful).

A SAMPLE STUDY – DISCRIMINATION OF MULTIPLE GROUPS

As an example of the first type of use mentioned above, let us consider again several of the variables in the sample study of the use of educational television to combat sex-role stereotyping in children's judgments of occupations.

Problem

Recall that in the earlier description of the educational television series (Chapter 13, pp. 162–167) children had been exposed to a series of television programs and classroom activities tended to reduce sex-role stereotyping. Subsequently, the children were tested on the degree to which they imposed male or female stereotypes when estimating who would fill a variety of occupations. This provided a male stereotype score (variable 2; Table 13.1) and a female one (variable 3). Also, the students answered multiple-choice test questions over the recall of television program content (variable 4) and understanding of instructional objectives (variable 6), As we have seen in examples of factor analysis and canonical correlation, there were differences in children's stereotyping and comprehension scores and these differences were variously related to programs they had seen, the number of classroom activities engaged in, and their teachers' attitudes toward the materials. For the present problem, the researcher is concerned with the degree to which the effects on stereotyping and comprehension scores are generalizable across students from the fourth, fifth, and sixth grades who were involved in the study. There is reason to believe, for example, that the older students may be more fixed in their stereotyped attitudes than the younger ones, and may thus appear different in terms of the stereotype scores. Also, the producers of the television series and activities are concerned with the degree to which they successfully gauge the materials for children in this age group, having anticipated that most of what they presented was comprehensible to fourth through sixth graders. Finally, since the results of the project are in both the areas of attitudes and comprehension effects, it is important to see the degree to which these two variables together might be different among the three groups.

The research question is: To what degree and how might the stereotyping scores and comprehension scores discriminate among fourth, fifth, and sixth graders in the project?

Method

It may be recalled that in the factor analysis (Chapter 13, page 166), it was found useful to combine the male and female stereotyping scores into a single *stereotyping* variable and the measures of recall, comprehension, and reasoning into one *understanding* variable. These two overall variables are then entered as the *discriminating variables* in the analysis. The *groups variable* in the analysis is whether a student was in fourth, fifth, or sixth grade.

A multiple discriminant analysis is undertaken, and it yields two *discriminant functions.* (Generally, the maximum number of discriminant functions is one less than the number of groups being compared.) Each of the discriminant functions represents a mathematical attempt to maximize a linear combination of the discriminating variables with the group variables. Like canonical correlation, the first function represents the initial maximum linear configuration for discrimination of the dependent variable, then the second function in this case maximizes whatever is not included in the first function, and so on. For each function, calculations will yield: (1) a *canonical correlation coefficient* describing the relationship of the discriminating variables with the groups variable; (2) the *standardized discriminant function coefficients,* which describe in standard score form the weightings of each of the discriminant variables on each of the functions; (3) the *unstandardized discriminant function coefficients,* which allow us to calculate a discriminant score by application to our raw measures; and (4) the *centroids,* or essentially the mean discriminant scores for each group on each function.[1]

Results

For the sample problem, the multiple discriminant calculations reveal that the canonical correlation between the first discriminant function and the groups is +.728. This is a moderately high correlation and statistically significant ($p < .01$). By contrast, the canonical correlation between the second discriminant function and the groups is +.143, which is low and not significant. Because of this we shall drop the second discriminant function from further practical interpretation. We know that the groups are discriminated chiefly on the first function.

What is the makeup of the first function? The standardized discriminant function coefficients, which tell us this, are nearly one for the *understanding* variable, and nearly zero for the *stereotyping* one. The interpretation is that, for all practical purposes, the first discriminant function mainly reflects scores on the understanding variable. Thus the answer to the basic question is that the fourth-, fifth-, and sixth-grade groups are significantly discriminated on the understanding measures and not on the stereotyping ones.

Unstandardized discriminant function coefficients allow one to take raw scores on the discriminating variables, plus a constant, and convert them into discriminant scores. (If we first standardize our measures, then the standardized discriminant function coefficients may be used for this purpose.) Discriminant scores are standardized with the mean of all data across the groups equal to zero, and the standard deviation equal to 1.0. This allows us to interpret the distribution of discriminant scores in terms of probabilities. In the present results the unstandardized coefficients are 0.58986 for the understanding score, 0.02519 for the stereotyping one, and −2.70259 for the constant. Thus, for example, if the mean raw scores of the fourth-grade group were 2.75 on understanding and 5.25 on stereotyping, these multiplied by their respective unstandardized coefficients and summed with the constant equal −0.95259, or the discriminant score for that group on the first discriminant function.

Recall that a *centroid* is the mean discriminant score on a function for a group. The mean discriminant score that we just calculated for the fourth graders is a centroid. For the other groups the centroids

---

[1] Other results are provided by computer programs for multiple discriminant analysis, such as Wilk's *lambda* calculations used to determine if amounts of variance accounted for by discriminant variables are significant, as well as several other calculations that are beyond the scope of this discussion.

on the first function are 0.13633 for fifth graders, and 0.81190 for sixth graders. From these we can see that understanding increases across the grades.

We can also use multiple discriminant analysis results to compare a new case with prior classifications. Suppose, for example, that a new fifth-grade class went through the same study and its understanding score was 4.8 and its stereotyping score was 5.0. Based on prior results, would they fit in with the other fifth graders? We can see by multiplying these raw scores by their respective unstandardized coefficients, then summing the results with the constant to yield a discriminant score. The calculated score is +0.25741, which is closest to the centroid of the earlier fifth graders (+0.13633); hence the new fifth graders would be classified into the same grade in terms of their performances.

Let us, however, look at another sample study where classification is the main objective.

A SAMPLE STUDY – USE IN CLASSIFICATION

As mentioned at the outset of this chapter, a second use of multiple discriminant analysis is in assigning previously unclassified individuals or phenomena to groups, based on results of a multiple discriminant analysis where both group assignment and the relation with the discriminant variables is known. This second sample study is also drawn from the earlier educational television sample study.

Problem

As previously described in Chapters 13 and 14, we know that children's understanding of the television materials devoted to reduction of sex-role stereotyping was predictable on the basis of the number of television programs they had viewed and the number of classroom activities in which they had engaged on the topic. For the present example, let us consider that an overall score of 4.0 or greater represents a *suitable level* of understanding of the materials, whereas a score lower than this represents an *unsuitable level*. Given the data we already have on children, it is possible to undertake a multiple discriminant analysis of the relationship between discriminant variables of television program viewing frequency and activities frequency with the group classification of having a suitable or unsuitable level of understanding of the materials. Results of the analysis could be a basis for classifying children into likely (1) suitable or (2) unsuitable comprehension groups based on a knowledge of the number of television programs they have viewed and activities they have engaged in, both as relevant as to the reduction of sex-role stereotyping.

Method

A discriminant analysis is undertaken where the discriminating variables are television viewing frequency (variable 8, Table 13.1) and activities frequency (variable 9). The groups variable is the separation of children with a suitable as against unsuitable level of understanding of the materials.

Results

For the sample problem the multiple discrimination analysis reveals one discriminant function with an associated canonical correlation of +.774. This is a relatively high degree of correlation between the discriminating variables and the groups.

Further, the standardized discriminant function coefficients are .781 for television viewing frequency and .492 for the frequency of classroom activities. It is clear that both variables contribute to differentiating groups of children in terms of suitable as against unsuitable levels of understanding of the materials. However, television viewing frequency contributes substantially more than does the activities frequency.

The unstandardized discriminant coefficients are 0.41110 for television viewing frequency and 0.17496 for activity frequency with a constant of −2.76479. If the mean television viewing frequency of the group having an *unsuitable* level of understanding is 3.40 and the activities mean is 2.60, then (applying the above) its centroid is −0.91217. Similarly, with a mean viewing frequency of 6.0 and activities mean of 5.4286, the centroid of the *suitable* level of understanding group is 0.65155. The midpoint between these two values is −0.12845.

Consider the continuum below for discriminant scores in this problem:

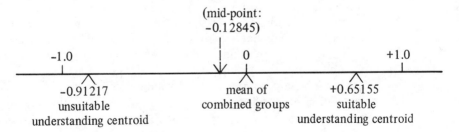

Again, the discriminant score is standardized; its mean equals 0, and this is the mean of the discriminant scores of both groups combined. Note relatively where the centroids of the two groups fall on this continuum. One practical way to classify new cases is on the proximity of their discriminant scores to the two group centroids. If, for example, a new child had a television viewing frequency of four programs and had undergone class activities four times, the discriminant score would be

$$(4)(.4111) + (4)(.17496) + (-2.76479)$$
$$\text{or}$$
$$-.42479$$

The new discriminant score is closer to the unsuitable understanding group centroid (to the left of -0.12845 above). This leads to the classification of the new case into that group.

NOTES ON MULTIPLE DISCRIMINANT ANALYSIS

Relation with Other Multivariate Methods

If you have read or even browsed in the earlier chapters on regression analysis, factor analysis, or canonical correlation, you should see a general conceptual relation of all of these methods with multiple discriminant analysis. As we said at the outset of this chapter, when the discrimination involves only two groups, multiple discriminant analysis and multiple regression analysis are essentially the same thing, except that the multiple discriminant analysis has a binary or dummy dependent variable in this case. In linear regression we are asking: What is the maximum combined linear combination of a cluster of variables with a dependent variable? In canonical correlation we are asking: What is the maximum linear relationship, or relationships, between two clusters of variables? In discriminant analysis we are asking: What is the maximum linear relationship (or relationships) of a cluster of variables with a variable that is divided into categories?

As in all the statistical methods ranging from analysis of variance through correlation, regression, and factor analysis, we are trying to account for variance reflected in relationships among variables. In multiple discriminant analysis, as is factor analysis and canonical correlation, we are taking the known relationships among individual variables and transforming them to another configuration. In factor analysis the larger array of relationships is reduced to fewer clusters of relationships called *factors*. In canonical correlation, clusters are interrelated in terms of *canonical variates*. In multiple discriminant analysis the relationship is characterized in *discriminant functions*. As in regression, factor analysis, or canonical correlation, we can attempt to interpret these clusters in terms of coefficients that indicate the degree of relationship of each of our original variables with the cluster. In the case of multiple discriminant analysis, each discriminant function is interpreted in terms of the various weightings of the original variables with that function.

Although the aforementioned methods all have basically much in common, they all have their own biases in what they reveal for us. The factors in factor analysis will, for the most part, be biased toward taking into account as much of the common variance in an overall matrix of interrelationships as is possible under the terms of the particular solution. In canonical correlation the canonical variates will be biased in definition toward the maximum linear relationships of two clusters of variables. As we said in the previous chapter, although the researcher can always see some similarities between canonical variates and factors for the same set of data, they will not necessarily be identical. A discriminant function, like a canonical variate or a factor, may reflect different aspects of the same patterns of common variance among an array of variables but will not necessarily have identical patterns with them.

Interpretation of Coefficients

Again, both standardized and unstandardized discriminant function coefficients are useful in interpretation of discriminant functions as well as discriminant scores.

Standardized coefficients, because they equalize measurement scales, are the basis for interpreting relative contributions of discriminant variables to a function. They can also be used to convert discriminating variable measures to discriminant scores *if* the former are first standardized (which they usually are not).

Unstandardized coefficients and the constant, like regression coefficients, are a basis for calculating discriminant scores from raw (unstandardized) scores of the discriminating variables. Because they also reflect measurement-scale differences, they are not an accurate basis for assessing relative contribution of discriminating variables to a discriminant function.

Particularly when multiple discriminate analysis is being used to classify new cases, based on known relationships between discriminating variables and groups, the equation involving unstandardized coefficients can be used in practical ways to show the overall relation as well as a practical guide for making predictions. The classification chart in Table 15.1 illustrates this for the second sample study.

Given the number of television programs viewed and the activities frequency as a basis for entering the chart, we can apply the coefficients and calculate likely group classification (S = satisfactory understanding, U = unsatisfactory). Note how the overall pattern in Table 15.1 reflects that television program exposure contributes

Table 15.1   **A Classification Guide for Predicting Likely Satisfactory (S) or Unsatisfactory Understanding (U) of Sex-Role Stereotyping Materials Given Frequency of Television Program Exposure and Class Activities (Simulated Data)**

TV Program Exposure	0	1	2	3	4	5	6	7	8	9	10	11	12	13	14	15	16
7	S	S	S	S	S	S	S	S									
6	U	S	S	S	S	S	S	S									
5	U	U	U	U	S	S	S	S									
4	U	U	U	U	U	U	U	S	S	S	S	S	S	S	S		
3	U	U	U	U	U	U	U	U	U	S	S	S	S	S	S		
2	U	U	U	U	U	U	U	U	U	U	U	U	S	S	S	S	S
1	U	U	U	U	U	U	U	U	U	U	U	U	U	U	S	S	S
0	U	U	U	U	U	U	U	U	U	U	U	U	U	U	U	U	S

Activities Frequency

more to gaining satisfactory understanding (S) than does activities frequency. If we were using this table for practical scheduling of programs, we would know that scheduling fewer than seven programs without activities would be unsatisfactory. Perhaps optimal is the combination of five programs and four or five activities sessions. Again, this table is one way the relations between the discriminating variables and the groups, as defined by the unstandardized coefficients, can be illustrated.

Discriminant Scores and Probabilities

It is useful to remember that these are standardized scores with a mean of zero and a standard deviation of one. The zero is the standard score representation of the overall mean of the combined groups on the discriminant score on that function. Thus if a centroid of one group is −1.0, we can estimate that 34 percent of the cases in the overall distribution are apt to fall between the mean of that group and the mean of the overall distribution. If the centroid of a second group is +0.50, then we know that 53 percent of the cases are likely to fall *between* the means of the two groups on that discriminant function. Or for that matter, if cases in a group range between −1.8 and −0.05 on a function, then 44 percent of the cases in the overall distribution are likely to fall in that group on that function.

Obviously when the correlation of a discriminant function is not perfect (that is, canonical correlation $\neq$ 1.0), this means that some error will be reflected in the assignment of cases to groups. This can be considered error in the body of data under analysis, or likely error in assigning new cases. In essence, "error" is when a case that prior data shows belonging to one group is, on the basis of the equation, assigned to another. Error is also reflected in the "overlap" of likely distributions around different centroids; the higher the proba-

Table 15.2   Classification of Students into "Suitable" and "Unsuitable Understanding" Groups Based on Multiple Discriminant Analysis of Television Viewing and Activities Frequencies (Simulated Data)

Known Group	Predicted Group	
	Unsatisfactory Understanding	Satisfactory Understanding
Unsatisfactory $(N = 50)$ understanding	50	0
Satisfactory $(N = 70)$ understanding	10	60

bility of overlap, the higher the likely error. Packaged computer programs for multiple discriminant analysis sometimes provide subroutines that will plot distributions and their overlaps. In some practical applications of multiple discriminant analysis to make classifications, the researcher may wish to designate a zone of "unclassified" cases where the probability of erroneous classification is high (say, as shown in a standard score plot).

It is also possible to calculate the assignment of cases to groups, based on the unstandardized coefficients, then to compare this with the known group assignment in the input data. This is illustrated in Table 15.2 for the second sample study. Note how more cases in the "satisfactory" understanding group are likely to be misclassified. This is because the distribution of this group overlaps more into the lower group rather than vice versa. (This was illustrated somewhat in the continuum on p. 189. Note how the centroid of the "satisfactory" group is closer to the overall mean or zero, and how the dividing point is to the left of this overall mean). Of the 120 cases, 110 or approximately 92 percent are assigned correctly. This is a very practical way of considering "error," both in the known data or in likely classification of new cases.

SUMMARY

Multiple discriminant analysis provides for the differentiation of single-variable groups or categories on the basis of relations with an array of discriminating variables. The calculation seeks to identify maximum multiple linear relations of discriminating variables with the groups. Each set of relations is called a *discriminant function*. For each discriminant function, the calculations also provide: (1) The canonical correlation of that function with the groups; (2) *standardized discriminant functions coefficients*, which reflect the relative contributions of discriminating variables to that function; (3) *unstandardized* discriminant function coefficients and a constant, which may be used to convert raw scores on discriminating variables to a discriminant function score; and (4) *centroids,* or the average discriminant function scores for each of the groups. Multiple discriminant analysis is used both to describe the differentiation of groups, based on discriminating variables, as well as a basis for classifying new cases into likely groups. Such analyses are part of a family of multivariate methods, all of which involve assessments of relations within, among, or between groups of variables. Discriminant function scores are standardized and are thus interpretable in terms of distributional probabilities. A practical gauge of error in multiple discriminant analysis is the comparison of prediction of group assignments with known ones.

SUPPLEMENTARY READINGS

Cooley, W. W., and P. R. Lohnes, *Multivariate Data Analysis.* New York: Wiley, 1971. Good overall coverage, but technical.

Kerlinger, Fred N., and E. J. Pedhazur, *Multiple Regression in Behavioral Research.* New York: Holt, Rinehart and Winston, 1973. Chapter 12 contains a brief, very clear, practical overview of multiple discriminant analysis.

Monge, Peter, and J. Cappella (Eds.), *Multivariate Techniques in Communication Research.* New York: Academic Press, 1978.

Nie, Norman H., C. Handlai Hull, Jean G. Jenkins, Karin Steinbrenner, and Dale H. Bent, *Statistical Packages for the Social Sciences.* New York: McGraw-Hill, 1975. Chapter 23 is a practical guide for a multiple discriminant analysis computer program.

TABLES

A. PERCENTAGE POINTS OF THE t DISTRIBUTION
B. PERCENTAGE POINTS OF THE F DISTRIBUTION
C. PERCENTAGE POINTS OF THE χ^2 DISTRIBUTION

USE OF THE t TABLE

In Table A, v stands for degrees of freedom ($d.f.$). When testing the difference between the means of two independent groups (Chapter 6), $v = n_I + n_{II} - 2$; for single or matched groups, $v = n - 1$. Enter the row corresponding to the value of v for the problem and locate the value of t that was calculated. The column headings provide the associated probability values (p). Probability values for one-tailed tests are those listed for Q; two-tailed test probabilities are those listed for $2Q$.

In practice, t will usually fall between two columns. Customarily we are interested in whether t falls in a position indicating $p < .05$ or $p < .01$, or whatever the significance level might be. For example, for a two-tailed test, if $v = 14$ and $t = 2.150$, then probability could be reported as $p < .05$. Sometimes it is desirable to report the probability values whose columns bracket the t value. Thus in the present case we might report $.02 < p < .05$, which is the same as saying, "probability greater than .02 but less than .05." If the calculated value of t exceeds the value shown in the column on the far right in Table A, it would be customary to report $p < .005$ for a one-tailed test, or $p < .01$ for a two-tailed test.

TABLE A. PERCENTAGE POINTS OF THE t DISTRIBUTION

ν	$Q = 0.4$ $2Q = 0.8$	0.25 0.5	0.1 0.2	0.05 0.1	0.025 0.05	0.01 0.02	0.005 0.01
1	0.325	1.000	3.078	6.314	12.706	31.821	63.657
2	.289	0.816	1.886	2.920	4.303	6.965	9.925
3	.277	.765	1.638	2.353	3.182	4.541	5.841
4	.271	.741	1.533	2.132	2.776	3.747	4.604
5	0.267	0.727	1.476	2.015	2.571	3.365	4.032
6	.265	.718	1.440	1.943	2.447	3.143	3.707
7	.263	.711	1.415	1.895	2.365	2.998	3.499
8	.262	.706	1.397	1.860	2.306	2.896	3.355
9	.261	.703	1.383	1.833	2.262	2.821	3.250
10	0.260	0.700	1.372	1.812	2.228	2.764	3.169
11	.260	.697	1.363	1.796	2.201	2.718	3.106
12	.259	.695	1.356	1.782	2.179	2.681	3.055
13	.259	.694	1.350	1.771	2.160	2.650	3.012
14	.258	.692	1.345	1.761	2.145	2.624	2.977
15	0.258	0.691	1.341	1.753	2.131	2.602	2.947
16	.258	.690	1.337	1.746	2.120	2.583	2.921
17	.257	.689	1.333	1.740	2.110	2.567	2.898
18	.257	.688	1.330	1.734	2.101	2.552	2.878
19	.257	.688	1.328	1.729	2.093	2.539	2.861
20	0.257	0.687	1.325	1.725	2.086	2.528	2.845
21	.257	.686	1.323	1.721	2.080	2.518	2.831
22	.256	.686	1.321	1.717	2.074	2.508	2.819
23	.256	.685	1.319	1.714	2.069	2.500	2.807
24	.256	.685	1.318	1.711	2.064	2.492	2.797
25	0.256	0.684	1.316	1.708	2.060	2.485	2.787
26	.256	.684	1.315	1.706	2.056	2.479	2.779
27	.256	.684	1.314	1.703	2.052	2.473	2.771
28	.256	.683	1.313	1.701	2.048	2.467	2.763
29	.256	.683	1.311	1.699	2.045	2.462	2.756
30	0.256	0.683	1.310	1.697	2.042	2.457	2.750
40	.255	.681	1.303	1.684	2.021	2.423	2.704
60	.254	.679	1.296	1.671	2.000	2.390	2.660
120	.254	.677	1.289	1.658	1.980	2.358	2.617
∞	.253	.674	1.282	1.645	1.960	2.326	2.576

This table is abridged from Table 12 in the *Biometrika Tables for Statisticians*, Vol. 1, Third edition (1966), E. S. Pearson and H. O. Hartley (Eds.). Reproduced by permission of the Biometrika Trustees.

Columns headed Q = 0.1 and 0.01 in Table A are taken from Table III of Fisher and Yates: *Statistical Tables for Biological, Agricultural and Medical Research*, published by Longman Group Ltd., London. (Previously published by Oliver and Boyd, Edinburgh), and by permission of the authors and publishers.

USE OF THE F TABLE

In Table B, v_1 stands for degrees of freedom (*d.f.*) in the numerator of the *F* ratio, and v_2 stands for degrees of freedom in the denominator. Determination of degrees of freedom for *F* ratios is discussed in Chapters 7 and 8. Table B is divided into two sections: "Upper 5 percent points" refers to values of *F* for significance at the .05 level; "Upper 1 percent points" refers to values for the .01 level. Depending on what significance level has been set, one of these tables is selected.

The table is entered at the intersection of the row (v_2) and column (v_1) corresponding to the degrees of freedom for a given *F* ratio. If the calculated value of *F* is equal to or exceeds the tabular entry, then $p < .05$ or $p < .01$, depending on which section of the table is being used. For example, if the significance level has been set at .05, and $v_1 = 3$, $v_2 = 9$, we would interpret an *F* of 3.90 as statistically significant ($p < .05$).

TABLE B. PERCENTAGE POINTS OF THE F DISTRIBUTION
UPPER 5 PERCENT POINTS

v_1 / v_2	1	2	3	4	5	6	7	8	9	10	12	15	120	∞
1	161.4	199.5	215.7	224.6	230.2	234.0	236.8	238.9	240.5	241.9	243.9	245.9	253.3	254.3
2	18.51	19.00	19.16	19.25	19.30	19.33	19.35	19.37	19.38	19.40	19.41	19.43	19.49	19.50
3	10.13	9.55	9.28	9.12	9.01	8.94	8.89	8.85	8.81	8.79	8.74	8.70	8.55	8.53
4	7.71	6.94	6.59	6.39	6.26	6.16	6.09	6.04	6.00	5.96	5.91	5.86	5.66	5.63
5	6.61	5.79	5.41	5.19	5.05	4.95	4.88	4.82	4.77	4.74	4.68	4.62	4.40	4.36
6	5.99	5.14	4.76	4.53	4.39	4.28	4.21	4.15	4.10	4.06	4.00	3.94	3.70	3.67
7	5.59	4.74	4.35	4.12	3.97	3.87	3.79	3.73	3.68	3.64	3.57	3.51	3.27	3.23
8	5.32	4.46	4.07	3.84	3.69	3.58	3.50	3.44	3.39	3.35	3.28	3.22	2.97	2.93
9	5.12	4.26	3.86	3.63	3.48	3.37	3.29	3.23	3.18	3.14	3.07	3.01	2.75	2.71
10	4.96	4.10	3.71	3.48	3.33	3.22	3.14	3.07	3.02	2.98	2.91	2.85	2.58	2.54
11	4.84	3.98	3.59	3.36	3.20	3.09	3.01	2.95	2.90	2.85	2.79	2.72	2.45	2.40
12	4.75	3.89	3.49	3.26	3.11	3.00	2.91	2.85	2.80	2.75	2.69	2.62	2.34	2.30
13	4.67	3.81	3.41	3.18	3.03	2.92	2.83	2.77	2.71	2.67	2.60	2.53	2.25	2.21
14	4.60	3.74	3.34	3.11	2.96	2.85	2.76	2.70	2.65	2.60	2.53	2.46	2.18	2.13
15	4.54	3.68	3.29	3.06	2.90	2.79	2.71	2.64	2.59	2.54	2.48	2.40	2.11	2.07
16	4.49	3.63	3.24	3.01	2.85	2.74	2.66	2.59	2.54	2.49	2.42	2.35	2.06	2.01
17	4.45	3.59	3.20	2.96	2.81	2.70	2.61	2.55	2.49	2.45	2.38	2.31	2.01	1.96
18	4.41	3.55	3.16	2.93	2.77	2.66	2.58	2.51	2.46	2.41	2.34	2.27	1.97	1.92
19	4.38	3.52	3.13	2.90	2.74	2.63	2.54	2.48	2.42	2.38	2.31	2.23	1.93	1.88
20	4.35	3.49	3.10	2.87	2.71	2.60	2.51	2.45	2.39	2.35	2.28	2.20	1.90	1.84
21	4.32	3.47	3.07	2.84	2.68	2.57	2.49	2.42	2.37	2.32	2.25	2.18	1.87	1.81
22	4.30	3.44	3.05	2.82	2.66	2.55	2.46	2.40	2.34	2.30	2.23	2.15	1.84	1.78
23	4.28	3.42	3.03	2.80	2.64	2.53	2.44	2.37	2.32	2.27	2.20	2.13	1.81	1.76
24	4.26	3.40	3.01	2.78	2.62	2.51	2.42	2.36	2.30	2.25	2.18	2.11	1.79	1.73
25	4.24	3.39	2.99	2.76	2.60	2.49	2.40	2.34	2.28	2.24	2.16	2.09	1.77	1.71
26	4.23	3.37	2.98	2.74	2.59	2.47	2.39	2.32	2.27	2.22	2.15	2.07	1.75	1.69
27	4.21	3.35	2.96	2.73	2.57	2.46	2.37	2.31	2.25	2.20	2.13	2.06	1.73	1.67
28	4.20	3.34	2.95	2.71	2.56	2.45	2.36	2.29	2.24	2.19	2.12	2.04	1.71	1.65
29	4.18	3.33	2.93	2.70	2.55	2.43	2.35	2.28	2.22	2.18	2.10	2.03	1.70	1.64
30	4.17	3.32	2.92	2.69	2.53	2.42	2.33	2.27	2.21	2.16	2.09	2.01	1.68	1.62
40	4.08	3.23	2.84	2.61	2.45	2.34	2.25	2.18	2.12	2.08	2.00	1.92	1.58	1.51
60	4.00	3.15	2.76	2.53	2.37	2.25	2.17	2.10	2.04	1.99	1.92	1.84	1.47	1.39
120	3.92	3.07	2.68	2.45	2.29	2.17	2.09	2.02	1.96	1.91	1.83	1.75	1.35	1.25
∞	3.84	3.00	2.60	2.37	2.21	2.10	2.01	1.94	1.88	1.83	1.75	1.67	1.22	1.00

This table is abridged from Table 18 in the *Biometrika Tables for Statisticians*, Vol. I, Third edition (1966), E. S. Pearson and H. O. Hartley (Eds.). Reproduced by permission of the Biometrika Trustees.

TABLE B. (*Cont.*) PERCENTAGE POINTS OF THE *F* DISTRIBUTION
UPPER 1 PERCENT POINTS

ν_2 \ ν_1	1	2	3	4	5	6	7	8	9	10	12	15	120	∞
1	4052.	4999.5	5403.	5625.	5764.	5859.	5928.	5982.	6022.	6056.	6106.	6157.	6339.	6366.
2	98.50	99.00	99.17	99.25	99.30	99.33	99.36	99.37	99.39	99.40	99.42	99.43	99.49	99.50
3	34.12	30.82	29.46	28.71	28.24	27.91	27.67	27.49	27.35	27.23	27.05	26.87	26.22	26.13
4	21.20	18.00	16.69	15.98	15.52	15.21	14.98	14.80	14.66	14.55	14.37	14.20	13.56	13.46
5	16.26	13.27	12.06	11.39	10.97	10.67	10.46	10.29	10.16	10.05	9.89	9.72	9.11	9.02
6	13.75	10.92	9.78	9.15	8.75	8.47	8.26	8.10	7.98	7.87	7.72	7.56	6.97	6.88
7	12.25	9.55	8.45	7.85	7.46	7.19	6.99	6.84	6.72	6.62	6.47	6.31	5.74	5.65
8	11.26	8.65	7.59	7.01	6.63	6.37	6.18	6.03	5.91	5.81	5.67	5.52	4.95	4.86
9	10.56	8.02	6.99	6.42	6.06	5.80	5.61	5.47	5.35	5.26	5.11	4.96	4.40	4.31
10	10.04	7.56	6.55	5.99	5.64	5.39	5.20	5.06	4.94	4.85	4.71	4.56	4.00	3.91
11	9.65	7.21	6.22	5.67	5.32	5.07	4.89	4.74	4.63	4.54	4.40	4.25	3.69	3.60
12	9.33	6.93	5.95	5.41	5.06	4.82	4.64	4.50	4.39	4.30	4.16	4.01	3.45	3.36
13	9.07	6.70	5.74	5.21	4.86	4.62	4.44	4.30	4.19	4.10	3.96	3.82	3.25	3.17
14	8.86	6.51	5.56	5.04	4.69	4.46	4.28	4.14	4.03	3.94	3.80	3.66	3.09	3.00
15	8.68	6.36	5.42	4.89	4.56	4.32	4.14	4.00	3.89	3.80	3.67	3.52	2.96	2.87
16	8.53	6.23	5.29	4.77	4.44	4.20	4.03	3.89	3.78	3.69	3.55	3.41	2.84	2.75
17	8.40	6.11	5.18	4.67	4.34	4.10	3.93	3.79	3.68	3.59	3.46	3.31	2.75	2.65
18	8.29	6.01	5.09	4.58	4.25	4.01	3.84	3.71	3.60	3.51	3.37	3.23	2.66	2.57
19	8.18	5.93	5.01	4.50	4.17	3.94	3.77	3.63	3.52	3.43	3.30	3.15	2.58	2.49
20	8.10	5.85	4.94	4.43	4.10	3.87	3.70	3.56	3.46	3.37	3.23	3.09	2.52	2.42
21	8.02	5.78	4.87	4.37	4.04	3.81	3.64	3.51	3.40	3.31	3.17	3.03	2.46	2.36
22	7.95	5.72	4.82	4.31	3.99	3.76	3.59	3.45	3.35	3.26	3.12	2.98	2.40	2.31
23	7.88	5.66	4.76	4.26	3.94	3.71	3.54	3.41	3.30	3.21	3.07	2.93	2.35	2.26
24	7.82	5.61	4.72	4.22	3.90	3.67	3.50	3.36	3.26	3.17	3.03	2.89	2.31	2.21
25	7.77	5.57	4.68	4.18	3.85	3.63	3.46	3.32	3.22	3.13	2.99	2.85	2.27	2.17
26	7.72	5.53	4.64	4.14	3.82	3.59	3.42	3.29	3.18	3.09	2.96	2.81	2.23	2.13
27	7.68	5.49	4.60	4.11	3.78	3.56	3.39	3.26	3.15	3.06	2.93	2.78	2.20	2.10
28	7.64	5.45	4.57	4.07	3.75	3.53	3.36	3.23	3.12	3.03	2.90	2.75	2.17	2.06
29	7.60	5.42	4.54	4.04	3.73	3.50	3.33	3.20	3.09	3.00	2.87	2.73	2.14	2.03
30	7.56	5.39	4.51	4.02	3.70	3.47	3.30	3.17	3.07	2.98	2.84	2.70	2.11	2.01
40	7.31	5.18	4.31	3.83	3.51	3.29	3.12	2.99	2.89	2.80	2.66	2.52	1.92	1.80
60	7.08	4.98	4.13	3.65	3.34	3.12	2.95	2.82	2.72	2.63	2.50	2.35	1.73	1.60
120	6.85	4.79	3.95	3.48	3.17	2.96	2.79	2.66	2.56	2.47	2.34	2.19	1.53	1.38
∞	6.63	4.61	3.78	3.32	3.02	2.80	2.64	2.51	2.41	2.32	2.18	2.04	1.32	1.00

This table is abridged from Table 18 in the *Biometrika Tables for Statisticians,* Vol. I, Third edition (1966), E. S. Pearson and H. O. Hartley (Eds.). Reproduced by permission of the Biometrika Trustees.

USE OF THE χ^2 TABLE

In Table C, v stands for degrees of freedom (*d.f.*), and Q stands for probability values (*p*). As discussed in Chapter 9, the one-sample case of chi-square has v equal to the number of categories minus one; the multiple sample case has v equal to the number of rows minus one being compared times the number of columns minus one.

Enter the row corresponding to the value of v for the problem and locate the column or position between columns where the calculated value of χ^2 falls. Probability values are then read from the column headings. For example, if $v = 4$, and $\chi^2 = 13.50$, we would report $p < .01$, or $.005 < p < .01$. If χ^2 is a value greater than the value in the .001 column, it would be customary to report $p < .001$.

TABLE C. PERCENTAGE POINTS OF THE χ^2 DISTRIBUTION

ν \ Q	0.250	0.100	0.050	0.025	0.010	0.005	0.001
1	1.32330	2.70554	3.84146	5.02389	6.63490	7.87944	10.828
2	2.77259	4.60517	5.99147	7.37776	9.21034	10.5966	13.816
3	4.10835	6.25139	7.81473	9.34840	11.3449	12.8381	16.266
4	5.38527	7.77944	9.48773	11.1433	13.2767	14.8602	18.467
5	6.62568	9.23635	11.0705	12.8325	15.0863	16.7496	20.515
6	7.84080	10.6446	12.5916	14.4494	16.8119	18.5476	22.458
7	9.03715	12.0170	14.0671	16.0128	18.4753	20.2777	24.322
8	10.2188	13.3616	15.5073	17.5346	20.0902	21.9550	26.125
9	11.3887	14.6837	16.9190	19.0228	21.6660	23.5893	27.877
10	12.5489	15.9871	18.3070	20.4831	23.2093	25.1882	29.588
11	13.7007	17.2750	19.6751	21.9200	24.7250	26.7569	31.264
12	14.8454	18.5494	21.0261	23.3367	26.2170	28.2995	32.909
13	15.9839	19.8119	22.3621	24.7356	27.6883	29.8194	34.528
14	17.1170	21.0642	23.6848	26.1190	29.1413	31.3193	36.123
15	18.2451	22.3072	24.9958	27.4884	30.5779	32.8013	37.697
16	19.3688	23.5418	26.2962	28.8454	31.9999	34.2672	39.252
17	20.4887	24.7690	27.5871	30.1910	33.4087	35.7185	40.790
18	21.6049	25.9894	28.8693	31.5264	34.8053	37.1564	42.312
19	22.7178	27.2036	30.1435	32.8523	36.1908	38.5822	43.820
20	23.8277	28.4120	31.4104	34.1696	37.5662	39.9968	45.315
21	24.9348	29.6151	32.6705	35.4789	38.9321	41.4010	46.797
22	26.0393	30.8133	33.9244	36.7807	40.2894	42.7956	48.268
23	27.1413	32.0069	35.1725	38.0757	41.6384	44.1813	49.728
24	28.2412	33.1963	36.4151	39.3641	42.9798	45.5585	51.179
25	29.3389	34.3816	37.6525	40.6465	44.3141	46.9278	52.620
26	30.4345	35.5631	38.8852	41.9232	45.6417	48.2899	54.052
27	31.5284	36.7412	40.1133	43.1944	46.9630	49.6449	55.476
28	32.6205	37.9159	41.3372	44.4607	48.2782	50.9933	56.892
29	33.7109	39.0875	42.5569	45.7222	49.5879	52.3356	58.302
30	34.7998	40.2560	43.7729	46.9792	50.8922	53.6720	59.703
40	45.6160	51.8050	55.7585	59.3417	63.6907	66.7659	73.402
50	56.3336	63.1671	67.5048	71.4202	76.1539	79.4900	86.661
60	66.9814	74.3970	79.0819	83.2976	88.3794	91.9517	99.607
70	77.5766	85.5271	90.5312	95.0231	100.425	104.215	112.317
80	88.1303	96.5782	101.879	106.629	112.329	116.321	124.839
90	98.6499	107.565	113.145	118.136	124.116	128.299	137.208
100	109.141	118.498	124.342	129.561	135.807	140.169	149.449

This table is abridged from Table 8 in the *Biometrika Tables for Statisticians*, Vol. 1, Third edition (1966), E. S. Pearson and H. O. Hartley (Eds.). Reproduced by permission of the Biometrika Trustees.

INDEX